ELECTRICAL CIRCUIT VOLTAGE CMOS OPAMP DESIGN TECHNIQUES

WILLIAM COX

FOREWORD

To meet the growing demands of the present day communication system, faster and highly accurate Integrated Circuits (IC's) design pose challenging design specifications for the operational amplifiers. An operational amplifier is one of the most widely used functional block for high level analog and mixed mode design applications such as precise analog filters, Analog to Digital (A/D) and Digital to Analog (D/A) converters. Many circuits and systems have shown that their overall performances are conditioned by that of used operational amplifier, there by mainly it a design issues with both features of high gain and high bandwidth. Besides these, improving the performance of the fundamental amplifier structure, optimizing silicon area and increase in static power are the challenging task to improve system performance. Therefore, there is a strong demand for new or improved operational amplifier which providing higher performances.

Bulk MOSFET method is used for low voltage applications but it provide less gain and low output voltage swing. Other method is Two stage CMOS OTA. It provides high gain but occupies larger area and consumes more power, another drawback is stability. Therefore single stage OTAs are suitable for low voltage and high gain applications because it provide high gain, high bandwidth, high speed, high voltage swing and occupy less area with optimize power consumption.

In this thesis work, a low voltage and high gain single stage operational amplifier structures are presented. The idea is to achieve improved Gain Bandwidth Product (GBW), DC gain and slewrate. This is well known that, these parameters are important for high frequency, fast settling applications, such as Switched Capacitor Filters, and Analog to Digital converters. They have vast application areas through analog design field, implemented in various technologies like 0.3µm, 180nm and 100nm. The proposed Gain Boosted OTA achieves 1.8GHz (GBW), 110db (DC) gain and 20V/µs slewrate in 2.5V supply voltage. The static power consumption is 18mW while driving 0.4pF load capacitor. The application of Local Common Mode Feedback (LCMFB) to the conventional Operational Transconductance Amplifier (OTA) structure provides significant increase in gain and slewrate performance without an increase in high static power and limited additional silicon area. Both single ended and the fully differential amplifier architectures are widely used in the industry. Features like high gain, Unity Gain Frequency (UGF), slewrate and speed are the requirements of the present day calling for modification to the existing architectures. Fully differential amplifiers are used to provide additional signal swing and reduce harmonic distortion characteristics. In the thesis LCMFB techniques are applied to both single ended and fully differential OTA. It has been noted that LCMFB provides wide range programming of amplifier characteristics and increases the versatility of the amplifier structure.

Telescopic Operational Transconductance Amplifier architecture provides high gain, high bandwidth and speed but its output signal swing is low. To overcome these problem n-Bias and p-Bias biasing architectures are designed. Telescopic architecture along with biasing architectures is known as Gain Boosted Operational Transconductance Amplifier. This architecture provides high gain, high bandwidth and high output signal swing.

Above architectures are designed in 0.18µm technology using Cadence tool. The specifications of the designed architectures are given below. These are suitable for specifications of A/D Converter resolution of 10 bit with sampling rate 100MHz consisting input voltage range of 1Vp-p and Latency less than 5 clock cycles.

1. **Single Ended Conventional Architecture:** Single Ended Conventional Architecture when designed and simulated, the frequency response shows the unity gain bandwidth of about 11.5MHz with an open loop DC gain of 53db,

5.3V/μs slewrate, 0.48mA maximum output current with settling time of 200ns. These parameters are not enough for high frequency applications like A/D converters. This architecture dissipates static power of 4.91 mW, 15.9mV of Input offset voltage and the input noise at 10MHz is 20μV.

2. Single Ended Conventional with Local Common Mode Feedback Architecture: Single Ended Conventional with Local Common Mode Feedback Architecture is designed and simulated. The frequency response shows the unity gain bandwidth of about 41.6MHz with an open loop DC gain of 67.8db, slewrate is 20.7V/μs, 2.34mA maximum output current with a settling time of 33.7ns. This architecture dissipates static power of 4.98 mW. 15.9mV of Input offset voltage and an input noise at 10MHz is 14μV. We observe that there is clear improvement in the performance parameters compared to the conventional CMOS OTA at the cost of reduction in phase margin. Therefore it is evident that the performance parameters of LCMFB CMOS OTA are better compared to conventional CMOS OTA. But these parameters are not enough for high frequency applications.

3. Fully Differential Architecture: Fully Differential Architectures when designed and simulated provide an open loop gain of 62 dB, with a Gain bandwidth of 160 MHz, and maximum output current of 0.48mA. It provides lesser slew rate i.e. 15.1 V/μs, dissipates static power of 200 mW, 1.9mV of Input offset voltage and an input noise of 18μV at 10MHz with Phase margin of 67 degree. This architecture gain and GBW is also low to improve these parameters LCMFB is used.

4. Fully Differential Architecture with Local Common Mode Feedback Architecture: Fully Differential Architecture with Local Common Mode Feedback Architectures is designed and simulated to provide an open loop DC gain of 67 dB, 270 MHz of Gain bandwidth, 2.08mA maximum output current, lesser slewrate i.e. 9.92V/μs, dissipates static power of 4.91mW, 8.3mV of Input offset voltage (Vos), with an input noise at 10MHz is 15μV. Its gain margin (dB) and phase margin (degree) is 15.6 and 78.5 respectively. Gain and GBW parameters are slightly increased.

5. Telescopic OTA: Telescopic Operational Transconductance Operational Architecture is designed and simulated to provide an Open loop DC gain of 60.7db, 1.8GHz of Unity Gain Frequency (UGF), Phase Margin (PM) of 62^0 with an Output voltage swing of 1.5V (p-p). In this architecture gain is low this parameter can be improved by biasing.

6. Auxiliary n-Bias OTA: This circuit is used for Bias in Telescopic OTA Architecture. This architecture provides the Open loop DC gain of 56.1db, 733MHz of Unity Gain Frequency (UGF), Phase Margin (P M) of 60.1^0 and an Output voltage swing of 1V (p-p).

7. Auxiliary P-Bias OTA: This circuit is used for bias in Telescopic OTA Architecture. This architecture provides Open loop DC gain of 62.2db, 692MHz of Unity Gain Frequency (UGF), Phase Margin (PM) of 59.3^0 with an Output voltage swing of 1V (p-p).

8. Gain Boosted OTA: A Combination of Telescopic OTA, n-Bias and p-Bias becomes Gain boosted OTA. This architecture provides Open loop DC gain of 110db. Unity Gain Frequency (UGF) of 1.8GHz, Phase Margin (PM) of 62.4^0 and Output voltage swing of 1.576V (p-p). This architecture is used for 10 bit A/D application.

Gain Boosted Operational Transconductance Amplifier is more suitable for specified analog to digital converters.

CONTENTS

List of Figures

CHAPTER 1

LOW VOLTAGE AND HIGH GAIN STANDARD CMOS OPAMP DESIGN TECHNIQUES

1.1   Introduction

The evolution of Very Large Scale Integration (VLSI) has formulated specific applications where millions of transistors can be incorporated on a one die or chip. Present trend mixed signal implementations and applications of integrated circuits Complementary Metal Oxide Semiconductor (CMOS) technology has became the main stay in present technology. Mixed signal integrated circuit imparts high chip compactness and less power on the digital circuit, and a better mingle of analog signals on the analog circuit. The scaling down of transistor's dimensions continued relentlessly over the past few decades has pushed down the supply voltages as well. In the present day, designing of significant and efficiency analog circuits has become more asserting with the existing orientation going close to low source voltages, because of the lower drive available for transistors. The design tradeoffs of digital designs have been thoroughly investigated compared to analog design. It has been shown that in digital design the lower voltage supplies can be utilized to reach an optimum performance in terms of the energy delay products.

Designing significant and efficiency analog circuits is quite difficult, unlike digital circuits, which can be designed to operate very efficiently at low voltages, where as analog circuits fail at low voltages. The fundamental obstruct in these circuits is the design of op-amp. At high source voltages, there is a tradeoff between power, speed, bandwidth and gain between other significant parameters. These significant parameters present conflicting alternative for the operational amplifier architectures. Operational amplifier is adapted for many different uses. It is difficult to design necessary basic blocks in these circuit design. Depends upon the assessment of circuit output impendence they are being restricted into categories of two. They are un-

1

buffered operational amplifier and buffered operational amplifier. Un-buffered operational amplifiers are Operational Transconductance Amplifiers (OTA) which have capacitive load and buffered operational amplifiers are voltage op-amps which consist less output impedance.

The main parameters in the present day electronic designs are bandwidth, power consumption and supply voltage. For a long time, power consumption and operating voltage were of least priority for Integrated Circuit (IC) designs engineers whereas bandwidth of the chip was the top priority. With the emergence of smaller transistor geometry and faster chips, the concern about heat dissipation and lifetime of device has increased. Smaller chips have also led to small electronic devices that are battery-powered and portable. In order to improve operation of device lifetime and extend the battery life of portable devices, efforts must be made to lower power consumption of internal circuits.

Lower overall device power consumption, the circuit operating voltages and currents necessary to perform the on-chip functions must be reduced. Complementary Metal Oxide Semiconductor (CMOS) technology has emerged as the dominant technology for IC design, because this technology offers the lowest power delay product, is simple to implement on silicon, and consumes the less quantity of power. In the introduction, priority is given to the component besides explaining the CMOS OTA architectures.

The geometry of complementary metal oxide semiconductor devices has been scaled down exponentially with time, and the channel length of CMOS devices have decreased. When the channel length of device is scaled down, the lateral electric field within the device increases. This reduces the device's reliability. Gate oxide thickness is also reduced, increasing the vertical electric field and leading to reduced gate oxide reliability. Increased packing density of devices causes more power dissipation per unit area, an increase in ambient temperature, and less stable circuit performance.

All these reliability issues can be alleviated by reducing the operating voltage. Current analog complementary metal oxide semiconductor circuit topologies utilize a 3.3V power supply voltage. This 3.3V operating voltage increases complexity and power consumption in new high-speed and portable systems operating at 1.8V that need to utilize the high-speed data transfer capabilities of analog input and output structures. The system must need additional voltage sources to accommodate the 3.3V analog systems. The reduction of analog operation voltage to 1.8V removes the need for additional voltage supplies and decreases the power consumption of the system. This operation of voltage reduction requires the implementation of low power CMOS design methodologies to analog circuit structures. Presently Silicon processing technology is established for 1.8V digital structures, and is ready for implementation in analog circuit applications where 1.8V power supply operation is necessary.

1.2 LOW-POWER/LOW-VOLTAGE EFFECTS ON CMOS TECHNOLOGY

1.2.1 Introduction to CMOS

To understand the effects of low-power/low-voltage on complementary metal oxide semiconductor devices, the interpretation of how the each and every CMOS circuit functions is necessary. CMOS technology consists of N-channel devices and P-channel devices together to benefit from what each has to offer. CMOS technology is accepted for being very high efficient. This technology has become the elite for a particular less power and less voltage application. In terms of an enhancement-mode, N-channel CMOS device, N+ source and drain regions are embed into a slightly doped P type silicon substrate, with a thin SiO2 gate oxide separating a polysilicon gate contact from the apparent of the silicon (Figure1.1). The silicon substrate terminal is often connected to the source terminal to keep it from having a floating potential, but it can also be connected to either power or ground as necessary.

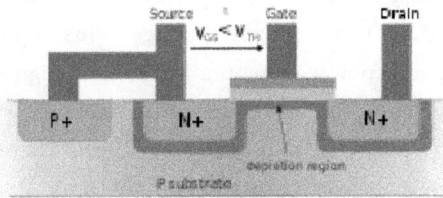

Fig.1.1 Diagram of N-channel CMOS Transistor

Parameters important to the design and operation of a complementary metal oxide semiconductor device include threshold voltage, capacitance, body effect, narrow channel effects, sub threshold current and short channel effects. All these parameters impact the operation of a CMOS device directly, and are effective when implemented with small geometry and low-voltage.

1.2.2 Introduction to Threshold Voltage

Complementary metal oxide semiconductor devices can be anticipation of as ON and OFF switches. The threshold voltage (Vt) is fundamentally the gate voltage (Vg) elementary to make a conducting channel in the CMOS device, or the voltage compulsory to make the switch ON. This definition has exceptions such as N-channel depletion mode transistors. In n-type channel depletion mode transistors, a channel already exists and the device is normally in the ON state when the applied gate voltage is zero. A negative gate voltage is needed to turn the device OFF. For normal or enhancement mode devices, before the gate voltage reaches the threshold voltage, the Metal Oxide Semiconductor (MOS) device is in the weak inversion region, and can be considered as OFF. Because no current is present among the source region and drain region. When the gate voltage is not positive enough to reach the threshold voltage of a long channel

4

Metal Oxide Semiconductor (MOS) device, the width of the depletion region under the gate is a operation of the gate voltage [3].

When a positive (+ve) voltage is related across the gate terminal relative to the source terminal of an NMOS device, negative charges are induced in the P type depletion region below the gate. A negative channel forms near the surface of this region, temporarily giving the area near the surface of the P-type material properties of N-type material. This phenomenon is known as inversion, and is the means for which MOS transistor operation can occur. Mobile electrons are able to travel between the source and drain during inversion (Figure 1.2).

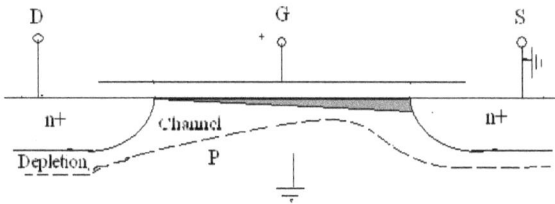

Fig.1.2 Diagram of Conducting Channel in an N-channel CMOS Transistor

Increasing the gate terminal voltage further leads to strong inversion. Strong inversion presents when the surface of the conducting channel is just as strongly N-type as the substrate is P-type. The length of the device depletion region continues to grow until strong inversion is reached. In strong inversion, increasing gate terminal voltage ceases to increase depletion width and increases only inversion. The voltage needed for strong inversion is known as the threshold voltage of the device. Once threshold voltage is reached, an increase in drain terminal voltage causes the channel to pinch off on the drain end of the MOS device and the drain current saturates. The device is now in

5

the saturation region of operation where the drain current (I$_D$) is essentially linear for any further increases in drain voltage. As Figure 1.3 depicts, for the applied gate voltage there is a drain voltage that causes the drain current of the device to become saturated [1].

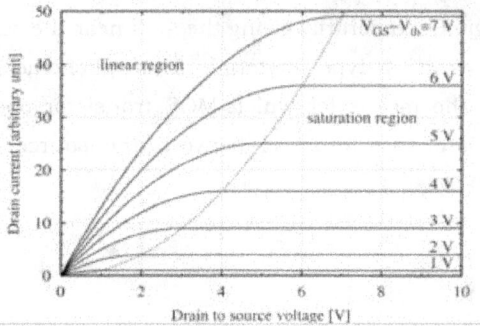

Fig.1.3 Diagram of Current Voltage Relationship for an N Channel CMOS Transistor

Similarly P type channel devices are made on an N type substrate with P+ implanted source and drain regions. A negative voltage along with the source voltage is given to the gate terminal, and positive charges (holes) are generates in the N-type depletion region. The channel between the source and drain a terminal are constructed and allows current conduction among the source terminal and drain terminal of the device.

1.2.3 Threshold Voltage and Effects of Real Devices

Solving Poisson equation in the device depletion region under the channel provides a simplified threshold voltage (V$_t$) formula. Considering

Gauss's law at the interface, since the displacement is continuous, the displacement at the oxide interface is equivalent to the overall space charge in the substrate depletion region [3]. The ideal threshold voltage required for strong inversion can be expressed as

$$V_T := \frac{-Q_d}{C_i} + 2\varphi_F \qquad (1.1)$$

Strong inversion in the depletion region at, where Q_d is the charge per unit area and C_i is the capacitance per unit area of the gate oxide insulator.

There are charges at the silicon and silicon dioxide (Si-SiO2) interface and within the oxide. Oxide charges are formed by contamination of sodium ions during processing [1]. Surface charges are formed at the sudden termination of the oxide layer where ionic silicon (Si) has been left at the surface. The ions and uncompleted silicon (Si) bonds form a sheet of positive charge at the interface. Sodium contamination can be controlled to a certain extent by using extremely clean materials in processing, and using {100} lattice arrangement wafers can minimize the amount surface charges at the Si-SiO2 interface [1, 10].

The change in the behavior of Metal Oxide Semiconductor (MOS) devices is enough that these effects need to be included in the equations. These changes are accounted by adding terms to the existing equations for the threshold voltage and other effected parameters. The ideal threshold voltage equation does not take into cause the sheet of positive charge at the Si-SiO2 interface or the differences in the Si-poly work functions, and assumes that the transistor is in the flat band condition before any applied voltage is put on the gate. This is not what actually happens in the device. To obtain a more accurate threshold voltage for the device the equation for the flat band voltage has to be appending to the existing threshold voltage equation.

$$V_T := \varphi_{ms} - \frac{Q_i}{C_i} - \frac{Q_d}{C_i} + 2\varphi_F \qquad (1.2)$$

Where φ_{ms} the work action of the metal-semiconductor is interface, and Q_i is the effective positive charge at the interface. The voltage required to create strong inversion must therefore be large enough to reach the flat band condition (φ_{ms} and $\frac{Q_i}{C_i}$), convenience the charge in the depletion region ($\frac{Q_d}{C_i}$), and lastly to induce the inverted region ($2\varphi_F$) [1]. This equation works for both N-type device and P-type device if appropriate signs are considered for each term.

.

All the terms in the new threshold voltage equation depend on doping except for $\frac{Q_i}{C_i}$ [1]. Figure1.4 describes how the level of doping affects the threshold voltage. As the amount of doping increases, so does the threshold voltage. Adjusting the amount of doping can control the level of the threshold voltage (V_t) to a certain extent.

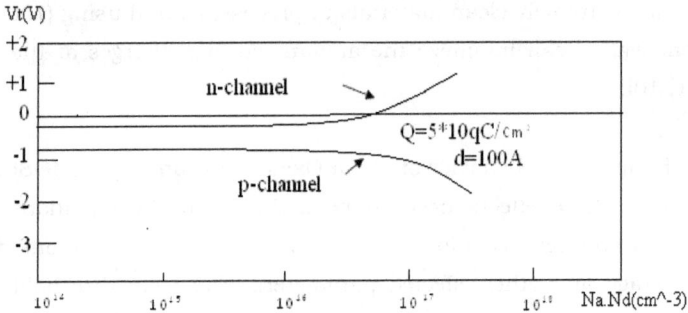

Fig.1.4 Doping Effects on threshold voltage

1.2.4 Body Effect

All Metal Oxide Semiconductor transistors have a fourth terminal which is called a bulk connection or back gate connection. This connection is almost always connected to one of the supply rails in integrated circuit designs. Fig.1.5. shows the bulk connection for an NMOS device as V_b.

8

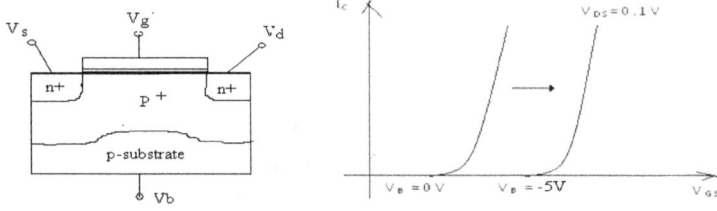

Fig.1.5 Bulk Connection for NMOS Transistor and Back Gate Bias Effect on Threshold Voltage of an NMOS Transistor

The threshold voltage (V_t) changes in the positive direction when the back gate bias changes from ground to a negative voltage. This is known as the body effect or back gate bias effect. It can be viewed as the variance in depletion width of the reverse-biased P-N junction among the source and the substrate due to the variance in the source bulk voltage [3]. The threshold voltage (V_t) equation must therefore be modified again to include the body effect.

$$V_T := V_{TO} + \gamma\left(\sqrt{V_{sb} - 2\varphi_F} - \sqrt{-2\varphi_F}\right) \qquad (1.3)$$

$$\gamma := \frac{\sqrt{2\epsilon_{si}.q.N_A}}{C_i} \qquad (1.4)$$

V_{TO} is the zero-biased threshold voltage equation and γ is the body effect coefficient. When designing circuits, the body effect coefficient should be as less as possible [4]. Lowering the substrate doping density and/or decreasing the thickness of the gate oxide both can reduce the level of the body effect coefficient [3]. In N-well complementary metal oxide semiconductor technology, the N-well doping density is higher than that of the substrate; therefore the body effect of the PMOS device is greater than that of the NMOS device. When scaling down geometry in complementary metal oxide semiconductor technology, the substrate doping density is increased and the gate oxide is thinner, resulting in mixed influences in the body effect coefficient [3].

9

1.2.5 Short Channel Effects

Due to geometric downsizing and advanced processing technology, Metal Oxide Semiconductor (MOS) devices with channel lengths of much less than one micron are possible. These so-called "short channel" devices are attractive because they reduce overall die area, can operate with less power, and can have a lower threshold voltage. The shortened channel extra effects from the source and drain regions that must be considered into account. The depletion region under the gate for a metal oxide semiconductor device with a long channel is shaped like a rectangle. The influence from the source region and drain regions can be disregarded. When depletion channel length is made small below a micron, the influence of the source region and drain region must be taken into account. The trapezoidal channel shape displayed in Figure 1.6 for an NMOS device is a direct result of the reduced proximity of the source region and drain region to one another.

The source and drain depletion regions intrusion into the device channel can cause punch-through current failures from drain and source depletion regions overlapping and shorting out the device channel. This will cause the device not to respond to any applied gate voltage control, and just act like a short. The second graph in Fig.1.6 depicts the channel length effect on the threshold voltage (Vt) of the NMOS device. As the length of a long channel device is decreased, the threshold voltage (V_t) stays constant until the length becomes very short. At this point, the threshold voltage (V_t) also begins to decrease.

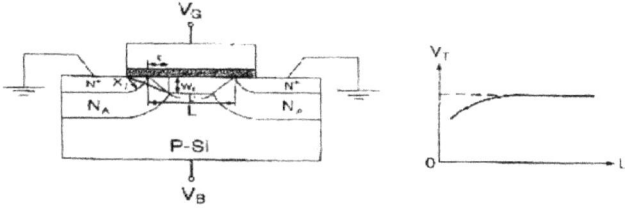

Fig.1.6 Short Channel NMOS Transistor and Effect of Length on Threshold
Voltage

The threshold voltage (V_T) equation must be altered to include these new
effects.

$$V_T := \varphi_{MS} - 2\varphi_F - \frac{Q_i}{C_i} - \frac{Q_d}{C_i}\left[1 - \left(\sqrt{1 + \frac{2W_d}{X_j}} - 1\right)\frac{X_j}{L}\right]$$ (1.5)

Above the equation, X_j known as the junction depth of the source/drain region,
L known as the length of the depletion region, and W_d known as the depth of
the depletion region. This equation is only applicable to devices with short
channel of less than one micron otherwise; the one-dimensional threshold
voltage equation is utilized. When devices are scaled down, X_j is reduced along
with a thinner gate oxide (sio2) to reduce the sensitivity of the threshold voltage
(V_T) to the effect of short channel.

Reduced geometry of short channel devices tends to increase electric
fields and cause effects from hot carriers to appear [1, 3]. In an NMOS device,
the field in the reverse biased drain junction can cause the carrier
multiplication and impact ionization. The resulting holes from carrier
multiplication contribute to the substrate current and can move to the source,
causing electrons to be injected into the p-region. Bipolar NPN transistor action
can result in the source-channel-drain configuration and cause the gate to lose
control of the current [3]. Transport of electrons through the barrier into the
gate oxide is another hot electron effect. These electrons can become taken in

11

the oxide and change the threshold voltage. These also affect the current-voltage characteristics of the device.

1.2.6 Narrow Channel Effects

Inverse of the short channel effects, as the device width decreases the threshold voltage (V_T) becomes more. This effect is generated by the influence of the depletion region below the edge of the field oxide surrounding the device active region and the channel stop implant under the field oxide surrounding the device active region [3]. This effect forces an addition to the short channel threshold voltage (V_T) equation.

$$V_T := \varphi_{MS} - 2\varphi_F - \frac{Q_i}{C_i} - \frac{Q_d}{C_1}\left[1 + \frac{\alpha W_d}{W} - \frac{X_J}{L}l + \frac{2\alpha W_d}{3W}\right] \cdot \sqrt{1 + \frac{2W_d}{X_J}l} \qquad (1.6)$$

Where α is between 0 and 1 and is calculated by the doping profile and topography of the field oxide at the Si-SiO_2 interface surrounding the active region [3].

1.2.7 Sub-Threshold Current

When an NMOS device is biased at $VG < VJ$, its drain current (I_D) is not really zero, but is instead exponentially proportional to the gate voltage [3]. The current leaking across the depletion region while the gate voltage (V_g) is less than the threshold voltage is called sub-threshold current. In weak inversion, the current in the depletion channel is caused by diffusion. With a thinner gate oxide (tox) and a more lightly doped substrate, the sub-threshold current slope is steeper. The metal oxide semiconductor device is more attractive for low power applications because it turns ON and OFF faster and has less leakage current. Figure 1.7 shows the sub-threshold current characteristics of an NMOS device with 10mV-drain voltage and bulk voltage plotted from 0V to -4V. As the bulk voltage decreases, the threshold voltage (V_t) increases and the sub threshold current decreases. This shows the tradeoffs between threshold voltage, sub threshold current and bulk voltage [3].

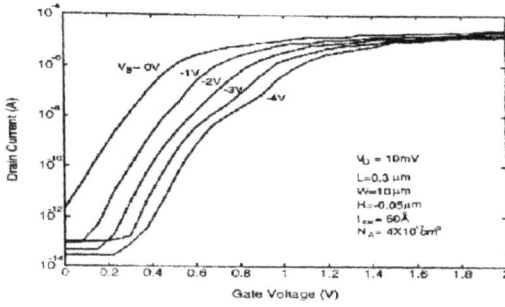

Fig.1.7 Effect of Back Gate Bias on Level of Sub-threshold Current

1.2.8 MOS Capacitances

For Metal Oxide Semiconductor (MOS) devices, there is a capacitance model that must be considered for AC and transient analysis. The capacitance model consists of two intrinsic capacitances; they are made up of the drain-gate capacitance and the source-gate capacitance.

The capacitance model consists of two extrinsic capacitances made up of the drain-body capacitance and the source-body capacitance [3]. The intrinsic performance of the metal oxide semiconductor device is most affected by the intrinsic capacitances. The source-gate and drain-gate capacitances are active and vary as a concern of the gate-source voltage [6]. The extrinsic capacitors act like parasitic capacitances, and have only secondary effects on overall device performance. Figure 1.8 shows the AC capacitance model for an NMOS device.

The drain-body capacitance and the source-body capacitance are parasitic capacitances among the source regions/drain regions and the bulk

13

device of the substrate. They occur because of the space charge in the depletion region of the reverse biased PN junctions [3].

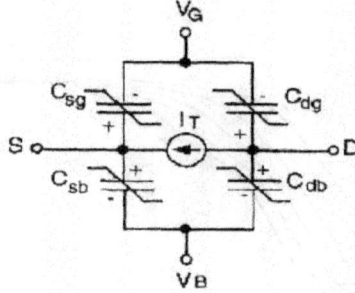

Fig.1.8 AC Capacitance Model for an NMOS Transistor

1.2.9 Low-Power/Low-Voltage Complications

As complementary metal oxide semiconductor devices shrink to smaller sizes, problems arise with increased currents and dissipation of power. When the length (L) and width (W) of complementary metal oxide semiconductor devices are reduced without reducing threshold voltage levels or power supply voltage, current increases, packing density of devices increases, and increases the amount of power dissipate. Increased power dissipation may cause an increased junction temperature and therefore an increased junction leakage current, making the device use more power than necessary [2]. Increased power dissipation in the die area has an adverse effect on the threshold voltage of the complementary metal oxide semiconductor devices. Threshold voltage exhibits a negative temperature coefficient, which causes enhancement mode devices with low thresholds to become depletion mode devices at elevated operating temperatures [7].

The solution to these problems is lowering of the source supply voltage. As the power supply voltage of the device is lowered, V_{GS}-V_{TH} lowers, and the drain current that is related to V_{GS}-V_{TH} also lowers. Propagation delay

14

increases, but the threshold voltage can be reduced to minimize the effect granted that an increase in sub-threshold current can be tolerated [8]. A phenomenon known as voltage bounce, related to the inductance of metal interconnect lines on the integrated circuits, may inhibit performance of low-voltage operation. As MOS device geometry shrinks, the lines get smaller in diameter and the inductance per unit length increases. When inductance is high, the change in power supply current (dl/dt) causes the voltage change on the lines by L (dl/dt).

The increased variation of delay in time associated with the change in threshold voltage after the supply voltage is reduced is another problem limiting the reduction of supply voltage. The propagation delay (t_{pd}) is a concern of the difference in threshold voltage (Vt) and supply voltage through the following equation.

$$\Delta t_{pd} \cong \left[\frac{\Delta V_T}{V_{DD} - V_T} \right] \lceil (r+1) \qquad 1 < r < 2 \qquad (1.7)$$

As the applied voltage is reduced, the change in the propagation delay (t_{pd}) due to the change in the threshold voltage may increase. This reduces circuit performance considerably, and increases the difficulty of the circuit design. To reduce this effect and increase performance stability, threshold voltage must again be cut down with retreat source supply voltage [9].

1.2.10 Supply Voltage Reduction Strategy

To lower the supply voltage for complementary metal oxide semiconductor devices, two strategies must be considered depending on the necessary system application [8, 9]. The high-performance strategy require lowering the supply voltage to increase system reliability including electro migration reliability, hot carrier reliability, oxide stress reliability, and other reliabilities related to the increased electric fields in the lower geometry devices. In this strategy, power supply voltage is not reduced aggressively and circuit performance is optimized. The low-power strategy results in degraded speed and circuit performance, but is more attractive for mobile systems in lengthening battery life. The reduction of the gate length does not automatically induce a reduced threshold voltage. Reducing the threshold voltage (V_{th}) can cause increases in levels of sub-threshold leakage current. This should be taken into consideration while lowering the threshold voltage of a sub-micron device.

The minimum threshold voltage can be determined through threshold voltage variation due to temperature effect, process fluctuation, and the ON-OFF current ratio of the device [3]. Figure 1.9 describes the threshold voltage versus supply voltage for various goals. The arrows describe which direction the supply voltage and threshold voltage should go for a particular goal. For example, if performance is to be increased, power supply voltage should be increased and threshold voltage should be decreased. By looking at this graph, three goals can be optimized into a point for the specified needs of the application - standby power, active power, and performance [9].

A decrease in the device power supply voltage is desired without more operating frequency loss and boosting power consumption. The design must take into account all of the low-power/low-voltage trade-offs and complications

16

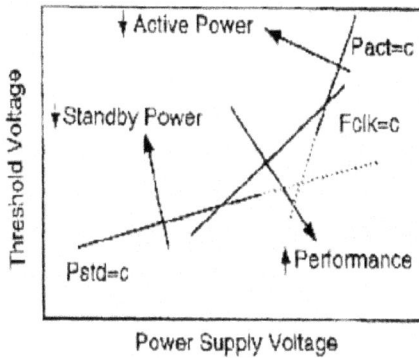

Fig.1.9 Trade Offs Associated with Power Consumption and Performance for Low-Power/Low-Voltage CMOS Design.

1.2.11 Operational Transconductance Amplifier

An operational transconductance amplifier (OTA) also known as a voltage controlled current source (VCCS) [6]. The OTA convert an applied voltage to an output current relative to a transconductance gain parameter $g_m = i_o/v_i$. An ideal transconductance amplifier characteristics are an infinite slewrate and bandwidth, along with an infinite input and output impendence $(Z_i = Z_o = \alpha)$ such that $i_i = i_{Ro} = 0$ and the output current is absorbed solely by the load.

Some of the key attractive properties of OTAs are their fast speed in comparison with conventional low output impedance operational amplifier and their bias dependence transconductance programmability. The main characteristics of practical operational transconductance amplifiers are limited linear input voltage range, finite output resistance, finite signal to noise (S/N) ratio and finite bandwidth. Using cascode structures output resistance can be improved high.

CMOS operational amplifier designs are available present days to drive only capacitive loads. With such capacitive loads, it is not compulsory to have a

17

voltage buffer to get less output resistance for the operational amplifier. As a conclusion, it is achievable to realize operational amplifiers consisting larger speeds and higher signal swing. These also must compel resistive loads.

These advancements are achieved by consisting only one high impedance node across the output of an operational amplifier that drives only capacitive loads. The admittance appeared across all nodes in the operational amplifiers is on the order of a transistors transconductance, and thus they consist approximately less resistances and impedances. Consist all circuit internal nodes of approximately less impedance; the speed of the operational amplifier is maximized. It should also be noted that these low node impedances significances in decreased voltage signals at internal every nodes other than output node.

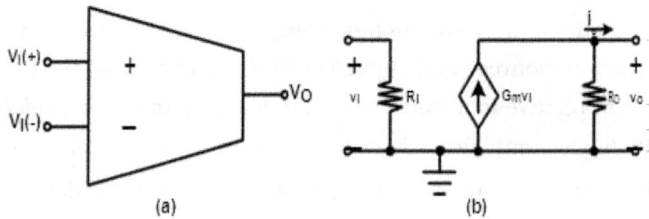

Fig.1.10 OTA Symbol and Equivalent circuit

The conventional operational transconductance amplifier is classified as a class A amplifier and is capable of generating maximum output currents equal to the bias current applied. The schematic symbol and equivalent circuit model for an operational transconductance amplifier is shown in Fig.1.10. Transconductance amplifier generates an output current (Io) equivalent to an applied voltage (vi) based on the transconductance gain (gm). Voltage gain of the open circuit Conventional OTA is given by $Av = g_m R_o$

1.2.12 OTAs with Single High Gain Stage

Operational transconductance amplifiers with a one (single) stage gain have been broadly used in applications of Switched Capacitors (SC). The gain boosting techniques can enhance large output resistance which present sufficient DC gain. Single stage OTA provides high bandwidth, phase margin and consumes small power.

Since the architecture is itself compensated, then not needed of frequency compensation i.e the load capacitance decides the dominant pole by which provides the footprint small on the die area. High circuit output resistance/impedance is accomplished by giving up the output voltage swing. The noise is more due to which the number of noise tenders devices.

1.2.13 Telescopic OTA

Telescopic operational transconductance amplifier represented in Fig.1.11 is the most accelerated desirable architecture. The gain bandwidth and the smallest non dominant pole are decided by the n-MOS devices, which results in stable phase margin and high bandwidth. The power consumption is low because the current legs being only less i.e. two. The disadvantage of single stage architecture is that it provides the narrow voltage swing at the output and input. The architecture gives output swing is limited of 2Vdsat less than V_{DD}. In the circuit another (low) side a minimum of 3Vdsat more than VSS. With this maximum possible circuit output swing and the circuit significant parameter input Common Mode (CM) range is zero. In general in circuits input CM range limited the circuit output swing. To enhance the DC gain an additional firm of cascodes can be added in together of the nMOS and pMOS surface. Applied supply voltage is Three volts or below, the swing of the output is not sufficient for A/D converter applications.

As shown in Fig.1.11, Telescopic opamp name came because the transistors are arranged one on the top of the other to establish a sort of telescopic composition. The current signals into common gate stages are injected by the architecture input differential pair. With a cascode current mirror the circuit then provides the differential applied voltage to single ended conversion. The parallel connections of two cascode configurations provide the low resistance of the signal and it provides large output node. So without limiting the circuit functionality the large resistance accounts the small signal gain.

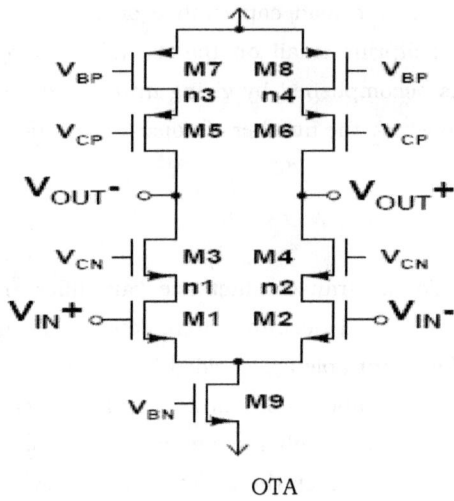

Fig.1.11 Telescopic OTA

Telescopic OTA advantages are it provides a gain similar to two stage OTAs architecture. Due to single pole Telescopic, OTAs are faster and it presents low noise factor. It provides good option for low power and less noise one stage OTAs. Telescopic OTA's disadvantage is that it provides the less dynamic range and less output swing.

1.2.14 Folded Cascode OTA

The Folded Cascode OTA represented in Figure 1.12. Switching circuits mostly prefer folded cascode operational amplifier architectures. Compared to telescopic OTA, Folded cascode OTA provides a high architecture applied input CM range and high output swing with the similar DC gain and besides not affecting the circuit speed. The circuit output swing, $V_{DD}-4Vdsat$, is not considering the input CM range, which is $V_{DD}-V_T-2Vdsat$ obtained using V_{GS} = V_T + Vdsat.

Based on the required phase margin, an n-MOS and p-MOS input pair has to be considered. The n-MOS input architecture, shown in Fig.1.12, provides high gain bandwidth (gm1/CL) due to the n-MOS input transistors, but the minimum non dominant pole (gm6/C1) correspondent with the node n1 is calculated by the less p-MOS transconductance and the high stray capacitances of the p-MOS current sources and the cascode devices. Considering a p-MOS circuit input pair provides less gain bandwidth, but the non-dominant pole is large, so credit to the n-MOS cascode device. Feed forward capacitors suggested to be considered to bypass the cascode transistors to enhance the phase margin at higher frequencies.

It is achievable to exploit an n-MOS and a p-MOS circuit input pair in parallel [4], it enhances the slewrate by 1/3 with the similar total current consumption. Due to this at the similar way time enhances the input capacitance. Thermal noise increases and it decreases the non dominant pole. Another achievable way to enhance the slewrate is that assure each transistors function in saturation region during deviate is to clamp the cascode nodes with diode connected devices.

In architecture the impedance appeared at the cascode node is really very high at the DC which is not discussed in many research papers. Operational amplifier output impendance is equal to folded cascode OTA dissipated by the gain of the cascode device (gm/gds). Thus, in the architecture impedance of the cascode node is presented in the order of 2/gds. At higher frequencies application the load capacitance make to change the architecture output impedance. Due to this circuit impedance available at cascode node decreases without calculating a low frequency pole in the frequency response.

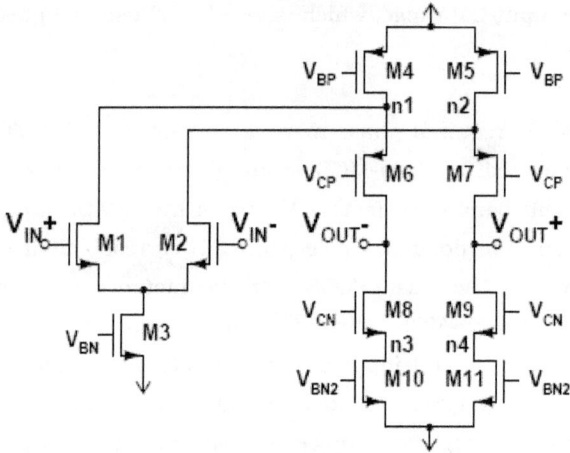

Fig.1.12 Folded cascode OTA

In analog to digital architectures high cascode impedance cannot be considered since it generate a Miller multiplication of the gate drain capacitance of the operational amplifier input device. The Miller effect can be avoided in the architecture adding cascode transistors on top of the circuit input pair. Added non dominant pole is greater than the one already available in the transfer function of n-MOS input pair and thus the phase margin is not decreased significantly. A different approch is to avoid the miller effect is to place capacitors, matched to the gate drain capacitance of the input device, the drain of the complementary input transistor and between the gates of the input transistor.

22

Compare to two stage operational amplifiers, Folded Cascode OTA has superior frequency response. Folded cascade OTA circuit provides better Power Supply Rejection Ratio (PSRR). In terms of Power consumption, both the circuits consume approximately equal power. It provides moderate voltage gain. Due to two extra current legs, there will be increase (i.e double) the power of dissipation. The folded cascode architecture has more devices, which accord significant input referred thermal noise to the signal. Compare to Telescopic OTA, this circuit has lower pole frequencies and less voltage gain.

1.2.15 Regulated Cascode (Gain Boosting)

Regulated cascode or Gain boosting architectures improve the gain without decreasing output voltage swing. It can be used in low voltage applications. Extra amplifier reduces the speed of the overall amplifier and also degrades the phase response/ phase margin. Due to pole-zero pairs this architecture leads instability.

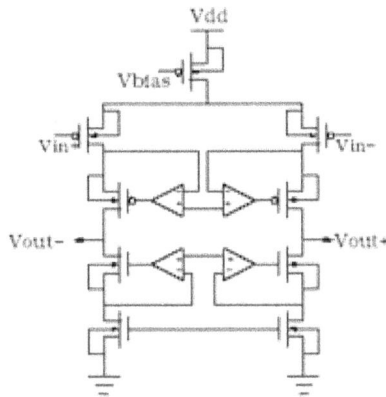

Fig.1.13 Regulated cascode (gain boost) OTA.

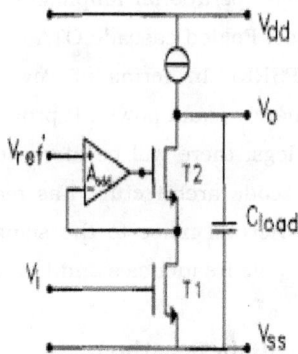

Fig.1.14 Basic Gain boost technique

In Fig. 1.14 shows the gain boost technique based on enhancing the cascoding effect of transistor $T2$ by including an additional gain stage. Due to this circuit output impedance is enhanced by the gain of the additional gain stage. Thus,

$$R_{out} = (g_{m2}r_{o2}(A_{add} + 1) + 1)r_{o1} + r_{o2} \qquad (1.8)$$

The DC gain can be enhanced by more orders of magnitude, is given as

$$A_{o,tot} = g_{m1}r_{o1}(g_{m2}r_{o2}(A_{add} + 1) + 1) \qquad (1.9)$$

In fig. 1.15shown the high frequency behavior of the gain increased cascade stage.

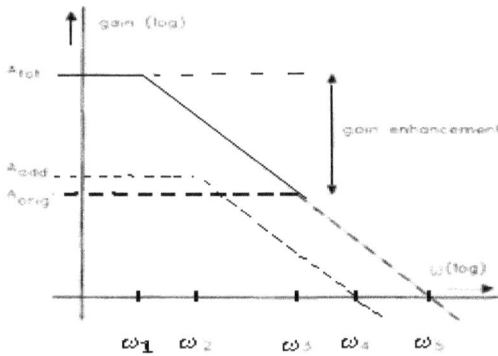

Fig.1.15 High frequency behavior of gain boost technique

Comparison of the architectures is shown below. In this design gain boosted amplifier has been chosen as the OTA architecture that meets the required specifications.

OTAs	Gain	Output Swing	Speed	Power Consumption	Noise
Telescopic Cascode	Medium	Medium	Highest	Low	Low
Folded Cascode	Medium	Medium	High	Medium	Medium
Two-Stage	High	Highest	Low	Medium	Low
Regulated Cascode (Gain	High	Medium	Medium	High	Medium

Boosting)					

Table 1.1 Performances of different OTAs

1.3 Thesis Organization

This dissertation presents design of low voltage and high gain complimentary operational amplifier techniques. Chapter 2 presents literature of Telescopic operational transconductance amplifier, Gain Boosted operational transconductance amplifier, Bulk driven MOSFET and Two stage operational transconductance amplifiers.

Chapter 3 describes a full analysis of the one-stage conventional operational transconductance amplifier structure. In this briefly explained about signal analysis i.e open loop gain, ac analysis, dc analysis, power dissipation, gain and phase margin. Apart from these fully differential structure also briefly explained.

Chapter 4 presents design of low voltage and high gain specifications. Design of operational transconductance amplifier with Local Common Mode Feedback described.

Chapter 5 described design methodology, spice models and PCB design. Chapter 6 presented results of designed Telescopic operational transconductance amplifier, folded cascode operational transconductance amplifier and Gain boosted operational transconductance amplifier. Comparison with various architectures also discussed. In this chapter presented application of analog to digital converter using above OTA. Chapter 7 provides conclusion and further scope of the study

1.4 Conclusion

Designing architectures with low voltage and high gain have become necessary for specific applications. In this chapter importance of threshold voltage, threshold voltage effects, body effects, short channel effects, narrow channel effects, sub threshold currents and MOS capacitance have been explained. In this section a brief explanation about Operational Transconductance Amplifier (OTA) architectures of Telescopic cascode, folded cascode and regulated cascode also have been included.

CHAPTER 2
REVIEW OF LITERATURE

2.1 Introduction

Designing the analog architectures which are suitable for low supply voltages is a tedious task. Accuracy and speed both are the very significant parameters of any analog circuit. It is challenging to optimize in the design of analog circuits and a compromise has to be reached between them. Optimize the architectures for these parameters supremacy to incompatible demands. In many analog circuits like Analog to Digital Converters, Switched Capacitor Filters, Sample and Hold circuits and Pipelined Analog to Digital Converters accuracy and speed both are determined by the settling behavior of the operational amplifiers. Fast and accurate settling needs more Unity Gain Frequency (UGF) of the amplifier and larger gain of the amplifier. A brief review is done for suitable circuits and described in this chapter.

2.2 Telescopic Operational Transconductance Amplifier

Sanchez-Sinencio.E et.al. [11] Presented an updated version of a 1985 tutorial paper on Complementary Metal Oxide Semiconductor (CMOS) transconductance amplifiers, active filters and different architectures. An Operational Transconductance Amplifier (OTA) known as the Voltage Controlled Current Source (VCCS). The maximum input voltage for a typical bipolar OTA is of the order of only 30mV, but with a transconductance gain tunability range of several decades. Since then, a number of researches have investigated ways to increase the input voltage range and to linearise the Operational Transconductance Amplifier (OTA).

Comparing with conventional low output impedance operational amplifiers, Operational Transconductance Amplifiers (OTAs) are high and their bias dependence transconductance is programmability. The wideband of the operational transconductance amplifier is due in part to the fact that their internal nodes are low impedances.

The author's view is that internal low impedance and parasitic capacitance still because a nonzero transconductance phase shift known as excess phase. When the Operational Transconductance Amplifiers (OTAs) are connected in a system in closed loop, the excess phase makes the actual frequency response deviate from the ideal case, especially for high Q systems. In the extreme case, the system may become unstable if the excess phase is not reduced.

Operational transconductance amplifiers main characteristics are i) Limited linear input voltage range ii) Finite bandwidth iii) Finite signal to noise ratio (S/N) and iv) Finite output impedance. According to the author S/N ratio is a function of the trans-conductance Operational Transconductance Amplifier (OTA). The architecture output impedance can be enhanced using cascode structures but we have to compromise on reduced output signal swing.

The Integrated Circuits pioneer works on transconductances using Bipolar Junction Transistors devices, Junction Field Effect Transistors devices and Complementary Metal Oxide Semiconductors (CMOS) devices were reported in the years 1980, 1981 and 1984 respectively. A number of significant contributions have been reported since 1985, including operational transconductance amplifiers for open loop applications such as continuous time filters, multipliers, non linear circuits and closed loop applications mainly for switched capacitor circuits.

Edgar Sanchez-Sinencio et.al. [12] Presented various transconductance amplifier topologies. An ideal transconductance amplifier is an infinite slew rate and infinite bandwidth. Ideally OTA consist an infinite input and output impedance. Fig.1.a. shows the simplest single input real transconductance. It is a metal oxide semiconductor (MOS) driver transistor M1 operating in saturation. One of the drawbacks of this architecture is low output impedance.

To increase output impedance several alternatives are shown in fig.2.d. It shows a group of cascode transconductors with high output impedance. In the Fig.2 b & c M1 operates in the ohmic region. This provides better linearity, but the transconductance is reduced in comparison with M1 operating in saturation. The amplifier A again increases the output resistance of the architecture shown in Fig.2.c.The folded cascode structure is shown in Fig.1.d.

Positive simple transconductor structure is shown in Fig.2.e. OTA symbol representation and equivalent model and Simple differential OTA shown in Fig.2.f and g, Balanced OTA shown in Fig.2.h, Conventional fully differential OTA without Common Mode Feedback (CMF) shown in Fig.2.i, Fully differential operational transconductance amplifier with inherent CMF shown in Fig.2. j, pseudo differential OTA shown in Fig.2. k.

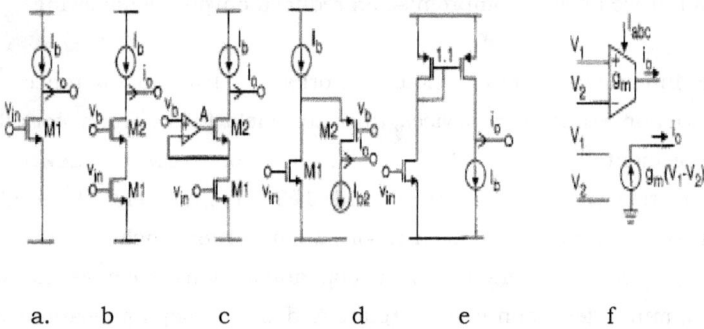

a.   b   c   d   e   f

g   h   i

j   k

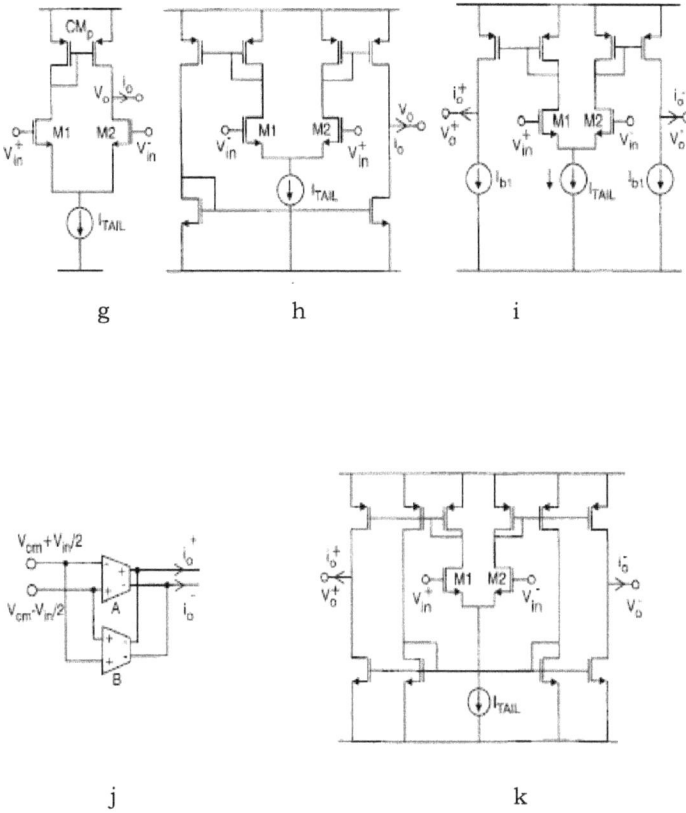

Fig.2.1. OTA Topologioes **a** Negative simple transconductor. **b**. Cascode transconductor. **c.** Enhanced transconductor. **d.** Folded cascode transconductor. **e.** Positive simple transconductor. **f.** OTA symbol representation and equivalent model. **g**. Simple differential OTA. **h**. Balanced OTA. **i**. Conventional fully differential OTA without CMF. **j.** Fully differential OTA with inherent CMF. **k**. Pseudo differential OTA.

31

Gursel Duzenli and Hakan kuntman [13] presented the necessity for bio telemetric implants, asks for MOS devices with high trans-conductance and reduced parasitic capacitance. Analog circuits with MOS transistor working in sub threshold region. Sub threshold region signifies a low voltage and a low power operation of analog devices with few nA. A MOS transistor working under sub threshold condition operates to a maximum of 1.5V. The applied low voltage gives flexibility to ulter the W/L ratio without affecting the electric field. To make the IC's cost effective scaling down of the device sizes plays an important role by facilitating more number of components integrated on a single chip.

Y.Taur [14] explained as we scale down dimensions of MOS devices the operating voltage range and the thickness of gate oxide should reduce. Generally for MOS devices, threshold-voltage, the supply voltage, and gate oxide thickness varies as a function of channel length. The scaling of devices has lead to reduction in length of channel however the supply voltage applied has not reduced proportionally.

K.Gulati and H.S.Lee [15] presented work in the area of Telescopic cascode operational amplifier. Higher operational frequency and less power consumption are the common characteristics of Telescopic opamp. Analog devices being responsive to common mode input and applied voltage variation in linear region is undesirable. The limitation of an output swing also is a major drawback. The paper presents some architectures of telescopic op-amp offering high output swing in comparison to the conventional. The new architectures exhibit higher common mode rejection ratio, power supply rejection ratio while keeping other performance parameters constant.

Gain enhancement and feedback circuit compensates for reduced common mode rejection ratio in the linear region due to low output resistance. The presented architecture in the paper showed an output swing of 0.7V which can be compared to the output swing obtained by conventional folded cascode op-amp.

Kush Gulati and Hac-Seunglee [16] presented the architecture of high performance Complementary Metal Oxide Semiconductor (CMOS) telescopic op-amp and exploration of the tail feedback in the architecture. Feedback in the architecture facilitates a high swing for the operational amplifier. A tail feedback technique compensates the degradation of differential gain and common mode rejection ratio.

J.Mulder et.al [17] proposed low voltage high swing cascode current mirrors. The paper presented designs of cascode current mirror and Wilson current mirror. These current mirrors are very useful in design of high speed converters.

A self biasing technique was introduced by **Feng Wang** [18] to present a classAB CMOS operational amplifier. The presented architecture suits for low voltage operations. The new architecture operational trans-conductance amplifier has improved characteristics compare to conventional operational trans-conductance amplifier. The OTA presented has a settling time independent of the slew rate and exhibits higher signal behavior. All in on the new architecture proposed provides good stability which is inherent to classAB operational amplifier.

Peter Bendix et.al [19] presented a paper on compact metal oxide semiconductor field effect transistor study of the relation between the structures and its ability to model harmonic distortion. In this paper explains problems in various MOSFET models and analyzes the problems to come to effective solutions.

Sohrab Emami et al. [20] presented a modified BSIM3V3 transistor model for a large wavelength signal (mm wave). A augmented BSIM3V3 model card was used to study the effect of parasitic capacitance on frequency signal. The DC curves relating to common source N-MOS transistor was matched to the core parameters of the BSIM3V3 to obtain the bias dependency of the small signal (mm wave). S parameters were measured and used to extract parasitic values of the frequency fit up to 65GHz.

Window of opportunity is presented for use of main stream complementary metal oxide semiconductor in mm waves. mm-Waves can only be possible, when 24GHz ISM and 60GHz band used in combination with CMOS scaling. The paper also demonstrates implementation of 0.18um CMOS operating at 24GHz for low noise amplifiers.

Boaz Shem-Tov, Miicahit Kozak and Eby G.Friedman [21] presented high speed CMOS operational amplifier. Design of the operational amplifier showed an output buffer following the operational transconductance amplifier. A component capacitor is put among input and output in of the circuit compensates the operational transconductance amplifier. A 0.18um CMOS technology was used in designing the architecture of operational amplifier.

For a load of 2pf and 1kohm connected par alley exhibited a dc gain of 86db, phase margin of 73 degree and unity gain frequency of 392MHz. The proposed method shows an increase of unity gain frequency by 1.5 times and an increase of 35 degree in phase margin for the given load condition. The presence of operational amplifier after the input in the direction of the output contributed greatly to the improvement of bandwidth and phase margin of the design architecture. Inverse relation between bandwidth and phase margin imposes a trade between speed and stability.

A novel technique proposed by **M.Di Ciano et al.** [22] discussed on use of current source as tail current for differential pairs to increase the output resistance of a metal oxide semiconductor. 0.35um CMOS technology along with SPICE simulation justifies the use of tail current. The obtained results showed gain of 2x (i.e 20db to 40db) while maintain the low supply voltage.

Girish Kurkure and Aloke K.Dutta. [23] Presented new adaptive biasing scheme for CMOS opamps. Adaptive biasing is utilized to optimize two contradicting circuit performance constraints to characteristics. They are slewrate limited period and the small signal settling period.

To increase speed, the slewrate of the opamp should be increased, which in turn requires that the amount of biasing current available to differential input stage of the opamp should be able to charge and discharge the load capacitor within the shortest possible time. Simple techniques of increasing the magnitude of this biasing current are detrimental due consequently to the large power dissipation problem.

To enhance the output of CMOS design **Burcin Serter Ergun and Hakan Kuntman** [24] introduced a novel realization technique on differential-output operational transconductance amplifier (DO-OTAs). Which provides higher linearity and circuit input signal range. But when this architecture is applied to other application the expected output resistance, input linearity range and circuit bandwidth was not sufficient.

In this architecture finite output made parallel to load capacitance (CL) due to which the circuit generated the filtering errors. This design is not suitable for high frequency applications. The circuits which enable filtering at low frequency and less output stage require very high output impedance architecture for such type of application this is applicable.

Yuzman Yusoff and Rohana [25] introduced a new architecture on high speed and high gain fully-differential (FD) Telescopic operational transconductance. This design is carried out in 0.18um CMOS technology. This architecture is suitable for 2.5 bit gain stage telescopic of 10 bit 50MS/S.

This specification can be considered for pipe line analog to digital converter. This design also provides 107db of gain with phase margin of 81 degree. This architecture results unit gain bandwidth of 763MHz and settling time of 8ns. The power supply of this design is 3.3V. This architecture consumes a very low power of 19Mw.

Tranwang Li, Ye Bo and Jiang Jinguang [26] presented a fully differential bandwidth enhanced telescopic operational transconductance amplifier in the paper. By recycling the tail current unity gain bandwidth is enhanced in the Telescopic OTA. This design work is carried out in 0.18um technology. This architecture results in 64 percentage enhancement of unity gain bandwidth. Tail current modified conventional telescopic operational amplifier explained in this paper shows an improved unity gain bandwidth.

2.3 Gain Boosted Operational Transconductance Amplifier

M.Rezaei et al. [27] proposed new architecture to improve the Slew Rate (SR) of the folded cascode operational transconductance amplifier. The proposed circuit is automatically activated during the slewing phase. Proposed architecture simulation results show an improvement by four folds in the slewrate and close to 40% reduction in the settling time as compared to a conventional folded amplifier.

High gain and high speed fully-Differential (FD) CMOS operational transconductance amplifier presented by **Zahra Haddad Derafshi and Mohammed Hossein Zarifi** [28] Which is used for low power applications. Gain boosting technique concept is introduced in the design so that this architecture can be utilized for portable low power consumption devices. This architecture is designed and simulated in 0.35um CMOS technology. This architecture results DC gain of 110db with phase margin of 71 degree. In this architecture Common Mode Feedback Circuit (CMFB) is used to make stabilization at high temperature condition.

It provides 800MHz of unity gain frequency and power dissipation of 5mw. This architecture is works at 3.3Volts power supply. These type of architectures are used for frequency selective filters where unwanted signals such as noise (ex. Thermal noise), interference and distortion products are separated from original message information. In communication applications these types of architectures are more preferred. Improved frequency response was obtained in simple Gm-C architecture presented in this paper.

HU, C [29] presented his views on limits of CMOS technology scaling and technologies beyond CMOS. He expressed his opinion on "Cramming more components onto Integrated Circuits", where Moore first proposed that transistor density on chips would grow exponentially. Since 1970's people have been predicting the end of CMOS. Despite these predictions, the monetary benefit of growth has driven people to do massive research, which has overcome all previous barriers. There are three key factors limiting continued scaling in CMOS. First is the minimum dimension that can be fabricated, second is diminishing returns in switching performance and the third off state leakage.

Novel technique Current Feedback Operational Transconductance Amplifier was presented by **G. Palmisano, G. Palumbo, and S. Pennisi** [30]. Two port theories contain four classes with respect to input terminal. Similarly four port theory contains nine classes with respect to input terminal. Common mode feedback operational transconductance (CFBOTA) belong to one of the nine classes. In this design available current can be utilized as CFBOTAs. This is more simple design used for current application devices.

Current Conveyor circuits were introduced by **Soliman A.Moahmood et al.** [31] in the paper. Apart from these important of current feedback operational amplifier is explained. Current Conveyor circuits are current mode operational amplifiers. This circuit provides high speed and larger bandwidth compared to voltage mode operational amplifiers. In mixed signal applications current mode operational amplifiers are preferred due to high speed characteristics.

Barath Kumar Thandri and Silva Martinez [32] demonstrated the Feed forward operational transconductance amplifier schemes. AMI 0.5mm CMOS process was used for fabrication. The paper compares feed forward OTA and conventional folded cascode OTA designs.

It is concluded that feed forward OTAs have better speed and more accuracy. These architectures are more suitable for Switched capacitor filters. Speed data conversions in mixed signal designs high resolution data converters having sampling frequency greater than 100MHz can be performed using these architectures. These architectures provide settling time compared to convention OTAs.

Beth Isaksen, Mike Holub, and James Griggs [33] presented a single-stage, fully differential amplifier with a gain of 80dB and a unity gain frequency of 1GHz for a 3V supply and 1pF load has been presented. A rail-to-rail CMFB scheme is used to achieve a differential output swing of 3.3V. The designed architecture uses a die area of 0.03mm² and consumes 17mW of power.

K.Bult and G.J.G.M.Geelen [34] presented single stage folded cascode operational transconductance amplifier. High DC gain and unity gain frequency can be calculated with this architecture. Architecture simulated and fabricated using 1.6um CMOS process. Architecture result shows 90db of DC gain and 116MHz of unity gain frequency with 16pf load capacitor (CL).

Single pole in the architecture leads to get fat settling time and high speed without affecting the output voltage swing. This architecture was operated at 5V power supply. Architecture provided 4.2V output swing without affecting the DC gain. 35pf capacitor load (CL) and band width of 18MHz was provided by the presented area. Single pole, architecture shows good stability and accuracy. In fast data converters these kinds of architectures are preferred. This architecture occupies less area on silicon die and consumes less power.

High output swing, high performance CMOS Telescopic operational transconductance amplifier described by **K.Gulati and H.S.Lee** [35], shows tail current and current source with transistors operating in deep linear region produced a high output swing. Common mode rejection ratio and differential gain of the designed architecture was not sufficient to use in low power applications. In this paper these drawbacks are overcome by new regulated cascode operational transconductance amplifier architecture. Both architectures are carried out in 0.18um CMOS technology.

Power supply provided for both architectures is 3.3V. Regulated cascode OTA resulted in differential swing of 2.15V. This architectures result shows

90db of differential gain and 90MHz of unity gain frequency. 50db common mode rejection ratio was obtained from the architecture. A large constant common mode rejection ratio was observed when simulated at high frequencies.

J.H. Hwang and C. Yoo [36] described a fully differential operational amplifier. Utilization of current reusing feed forward compensation scheme is presented in architecture. It resulted less power consumption and high bandwidth. In novel design architecture bias current is reused of the next stage of operational transconductance amplifier.

The architecture designed and simulated using 0.25um CMOS technology. Architecture result shows 77db of DC gain with phase margin of 56 degree. 870MHz of unity gain frequency achieved with load capacitance of 1pf. Architecture executed at 2.5V supply voltage and 1.8mA of current draws. Comparison of conventional OTA and feed-forward OTA along with compensation schemes explained with speed, bandwidth, and unity gain frequency and power consumption parameters.

A 16 bit pipeline analog to digital converter using operational transconductance was presented by **Nordiana Mukahar, Siti Hawa Ruslan and Warsuzarina Mat Jubadi** [37]. It is a challenging and difficult task to design analog circuits to work on less supply voltage at low frequency applications. 5V power supply voltage is applied for the architecture. Architecture designed and simulated using 0.5um CMOS14TB technology. The architecture provides 93.2db of gain, 9.32 MHz of unity gain frequency and 93.14 degree of phase with load capacitor 5pf.

2.4 Bulk Driven MOSFET

Jonathan Rosenfeld et.al, [38] Explained new concept of bulk driven MOSFET architecture. In this inversion layer is formed by applying gate to source voltage and bulk terminal is used for applying input signal. Signal path does not consist threshold voltage (Vt). Transconductance (gm) of the architecture is less than the gate transconductance by five times. Bulk driven or substrate driven MOSFETs provide very less gain due to low transconductance. 68db of gain is presented by this architecture. Advantage of such type of circuits is that it consumes very less power. These are suitable for bio medical applications.

J. Ramirez-Angulo et al. [39] described a 0.18um CMOS process for low voltage rail to rail fully-differential operational transconductance amplifier. Complementing input differential pairs of the circuit is used to get rail to rail operation. Guzink et al. was first introduced about the bulk driven MOS transistor in 1987. This paper also explained about auxiliary gain boosting amplifiers. This circuit provides 68db of open loop gain and operates at 0.8V power supply voltage. Its power consumption is 94mW.

A study on distortion analysis for non linear circuits in analog integrated circuits was carried out by **Yousuke Taniguchi et al.** [40]. A paper explains about frequency analysis of CMOS operational amplifiers at high frequencies. A large variation in output characteristics are observed due to the presence parasitic capacitances between source, drain, gate and substrate (bulk). A harmonic balance method is introduced to solve the frequency response curves. To carry out the Fourier transformation of the noise device with non linear parasitic elements a Fourier circuit's model was developed in SPICE using analog behavior models.

2.5 Two stage OTA

41

Rida Assadd and Jose Silva-Martinez [41] presented feed forward technique. In design of high frequency OTA feed forward techniques play an important role. In case of single stage amplifiers a recycling folded cascode operational transconductance amplifier shows a gain bandwidth of about ~200MHz as compared to 106MHz and a slew rate of 230V/μs against approximately 100V/μs for the conventional OTA. A multistage amplifier gives a high gain. No Capacitor Feed Forward (NCFF) technique having a high frequency pole zero doublets is employed to obtain a high gain.

Design architecture provides 90db of dc gain. GBW of 320MHz and 70 degree phase margin. The settling time is observed to be faster for the OTAs with NCFF topology as compared to miller compensation. To have a settling time of less than 4ns desirable in devices operating with sampling frequency more than 100MHz can be designed using this technique.

Along channel device biased at low current can use these cascode structures for high gain amplifiers while short channel devices are biased at high current levels for high bandwidth amplification. For high frequency applications a folded cascode OTA is preferred over the Telescopic OTA due to its higher signal swing inspire of parasitic pole and larger DC gain.

M. Yavari and O.Shoaei [42] presented a time domain model for the slewrate of CMOS two stage miller compensated OTA. Architecture simulations were done using HPSICE and 0.25mm CMOS technology.

RF circuts as low noise and power amplifiers was presented by **A.A.H.Ab-Rahman and I.Kamisian** [43]. The paper discusses mismatching threshold voltages between the neighboring MOSFETs. The difference in threshold voltages of two neighboring MOSFETs of the differential stage leads to a deviation in output.

Hamed Aminzadeh and Reza lotfi [44] presents a well defined procedure for the design of high speed two stage CMOS operational amplifiers. Cascode compensated amplifiers with good tradeoffs between speed, power and stability that make them suitable for high speed applications are discussed.

The architectures are analyzed to obtain the required circuit level parameters according to particular bandwidth and stability. The effect of new capacitor's sizing rules to split the compensation capacitance and increase the amplifiers speed has also been considered. Based on these novelties, a new way of the design of high speed operational amplifier is proposed.

Thomas Liechti et al. [45] describes the implementation of a 12bit 230MS/S pipelined ADC. Two stage folded cascode operational transconductance amplifier (OTA) architecture is used for improvements in settling performance.

This architecture is suitable for emerging high performance applications. This application asks for higher sampling rate in the order of 800MHz to 1GHz with higher bit resolution. These requirements can be achieved using time slotted multi channel ADC architecture.

2.6 Conclusion

In this chapter, literature of Telescopic OTA, Gain Boosted OTA, Bulk driven MOSFET and Two stage OTAs are reviewed. From literature survey, it is concluded that Bulk driven MOSFETS provide less gain. So, it is not suitable for low voltage and low gain applications and two stages OTA provides high gain but occupies larger area, consumes more power and has stability issues. Therefore Telescopic OTAs and Folded cascode OTAs are suitable for less voltage and high gain applications. Gain can be increased with gain boosting architectures.

CHAPTER 3

CONVENTIONAL OPERATIONAL TRANSCONDUCTANCE AMPLIFIER

3.1 Introduction

A conventional, one stage, Operational Transconductance Amplifier (OTA) configuration is shown in Fig.3.1.

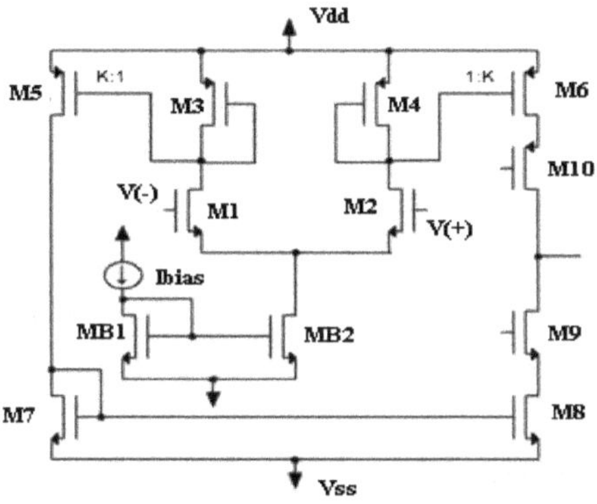

Fig.3.1 Conventional One Stage Operational Transconductance Amplifier

The Operational Transconductance Amplifier (OTA) employs a differential input pair and three current mirrors. The differential input pair comprises transistors M1 and M2. The differential pair is biased by MB1 and MB2. Mirrors formed by transistors M3, M5 and transistors M4, M6 reflect currents generated in the differential pair to the output. The current generated by the mirror of transistors M3 and M5 is then reflected to the output via the mirror formed by transistors M7 and M8. The mirror gain factor (K) indicates the gain in mirrors formed by transistors M3, M5 and M4, M6 with the relations $\beta_5 = K\beta_3$, $\beta_6 = K\beta_4$ where $\beta = KP/2(W/L)$. An increase in mirror gain factor will increase the slewrate and gain bandwidth of the conventional OTA. It increases area and static power dissipation and a decrease in phase margin. Cascoding transistors

45

in the circuit M9 and M10 are biased by V_{casn} and V_{casp}. It provides increased gain via increased (cascoded) output resistance

3.2 Operation

The conventional operational transconductance amplifier uses an input differential pair with three current mirrors to bias an applied input voltage into an output current. Input common mode signals ($V_i(+)=V_i(-)$) are, ideally rejected. For a common mode input voltage, the currents are constant in the circuit and will be $i_{d1}=i_{d2}=I_{BIAS}/2$, and output current is zero. A differential applied input signal will generate an output current proportional to the applied differential voltage based on the circuit transconductance of the differential pair.

The output stage is a push-pull structure, the conventional OTA is only capable of producing an output current with a maximum amplitude equal to the bias current in the output ($K^*I_{BIAS,OS}$). For this reason, the conventional OTA is referenced as a classA structure. It is capable of producing maximum signal currents equal to that of the bias current applied. Slewrate is defined as the maximum rate of change of the output voltage and it is directly proportional to the maximum output current. The slewrate can be calculated as the output current divided by the total load capacitance.

The conventional operational transcondctance amplifier therefore suffers the consequence that high speed requires large bias currents which translates to large static power dissipation. For many applications such as pipelined analog to digital converters and switched capacitors require high slewrate, high gain and high bandwidth with less static power dissipation. These parameters are difficult to get with classA structures. The proposed class AB structure with Local Common Mode Feedback (LCMFB) conventional operational transconductance amplifier is presented to achieve the above requirement.

3.3 Signal Analysis

3.3.1 Open Loop Gain

Fig.3.2 will be referenced to determine the open loop gain.

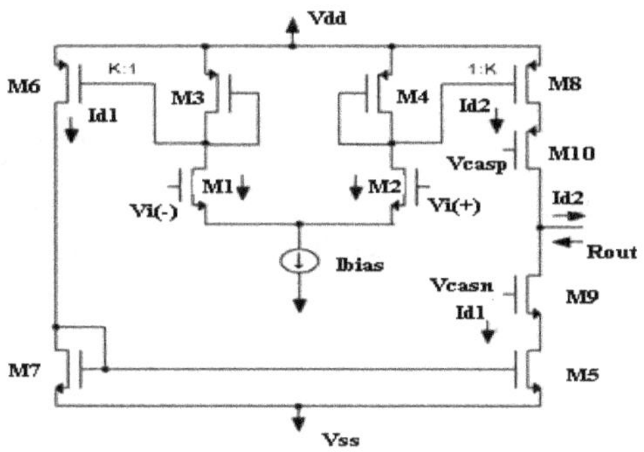

Fig.3.2 Conventional One Stage OTA Open Loop Gain Schematic

The output current, in terms of the mirror gain factor (K), is given by

$$i_{out} = Ki_{d\,2} - Ki_{d1} \qquad (3.1)$$

Where

$$i_{d1} = gm_1\,Vi(-), \quad i_{d\,2} = gm_2\,Vi(+) \qquad (3.2)$$

Assuming: $gm_1=gm_2$, and substituting (3.2) into (3.1)

$$i_{out} = Kgm_{1,2}\,(Vi(+) - Vi(-)) \qquad (3.3)$$

This indicates the transconductance gain of the OTA and is given by

$$Gm = Kgm_{1,2} \qquad (3.4)$$

The output resistance is a cascode resistance and is given by

$$R_{out} = gm_{10}r_{o10}r_{o6} \,//\, gm_9 r_{o9} r_{o8} \qquad (3.5)$$

Combining (3.3) and (3.5), the output voltage is then given by

$$v_{out} = i_{out}\,R_{out} = Kgm_{1,2}\,(Vi(+) - Vi(-))(gm_{10}r_{o10}r_{o6} \,//\, gm_9 r_{o9} r_{o8}) \qquad (3.6)$$

and the open loop gain is

$$A_{OL} = v_{out} / v_{in} = Kgm_{1,2}\,(gm_{10}r_{o10}r_{o6} \,//\, gm_9 r_{o9} r_{o8}) \qquad (3.7)$$

3.3.2 AC Analysis

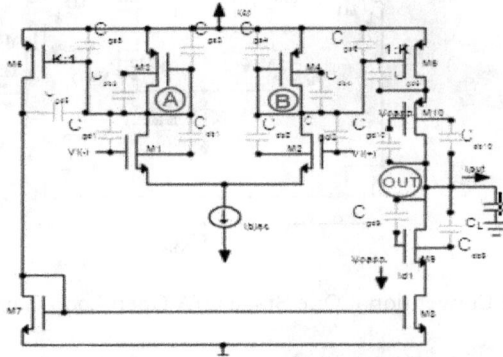

Fig.3.3 Conventional One Stage OTA AC Analysis Schematic

The gain bandwidth of the conventional one stage operational transconductance amplifier is limited mainly by the low impedance, high frequency, poles at nodes A/B, in conjunction with the high impedance, low frequency pole at the output node. The following analysis will define the high frequency pole and will assume that the nodes A and B are equivalent nodes in terms of resistance and parasitic capacitance (M1=M2, M3=M4, and M5=M6). The resistance at nodes A/B is dominated by the diode connected resistance $(1/gm)$ of M3, M4 and is given by

$$R_{A,B} = \frac{1}{gm_{3,4}} // r_{o1,2} \approx \frac{1}{gm_{3,4}} \qquad (3.8)$$

The parasitic capacitance at A/B is given by

$$C_{A,B} = C_{gd1,2} + C_{db1,2} + C_{db3,4} + C_{gs3,4} + C_{gd5,6} + C_{gs5,6} \approx C_{gs3,4} + C_{gs5,6} \qquad (3.9)$$

Combining (3.8) and (3.9), and the relation $C_{gs5,6} = KC_{gs3,4}$, the pole at A/B is

$$f_{pA,B} = \frac{1}{2\pi C_{A,B} R_{A,B}} = \frac{gm_{3,4}}{2\pi C_{GS3,4}(K+1)} \qquad (3.10)$$

The output node capacitance is dominated by the load capacitance (C_L).

$$C_{OUT} = C_{dg9} + C_{db9} + C_{dg10} + C_{db10} + C_L \approx C_L \qquad (3.11)$$

Combining (3.5) and (3.11), the dominant pole/bandwidth of the OTA is given by

$$f_{pOUT} = \frac{1}{2\pi C_{OUT} R_{OUT}} = \frac{1}{2\pi C_L (gm_{10} r_{o10} r_{o6} // gm_9 r_{o9} r_{o8})} = f_{3db} \qquad (3.12)$$

This analysis indicates the relation between the phase margin and the mirror gain factor. Equation (3.10) indicates the high frequency pole $f_{pA,B}$ is inversely proportional to mirror gain factor. An increase in mirror gain factor will result in a decrease in $f_{pA,B}$ and consequently a decrease in phase margin. The bandwidth of the conventional OTA is given in equation (3.12) and is inversely proportional to the load capacitance (C_L).

3.3.3 Gain Bandwidth (GB)

Equations (3.7) and (3.12) are combined for the gain bandwidth product.

$$GB = \frac{K gm_{1,2}}{2\pi C_L} \qquad (3.13)$$

3.3.4 Maximum Output Current

The maximum output current of the conventional OTA is limited by the mirror gain factor (K) and the bias current and is given by

$$I_{OUT}{}^{MAX} = K I_{BIAS} \qquad (3.14)$$

3.3.5 Slew Rate

The slewrate is given by

$$SR = \frac{I_{OUT}^{MAX}}{C_L} = \frac{KI_{BIAS}}{C_L} \qquad (3.15)$$

The slewrate therefore, increases linearly with K.

3.4 DC Analysis

3.4.1 Input Common Mode Range (CMR)

The common mode range is defined as the range of voltage (V_{IN}^{MAX}, V_{IN}^{MIN}) for which the input differential pair will remain in saturation. Common mode range is determined by the architecture, bias current and transistor sizes. For the differential input stage with diode connected loads, the minimum and maximum input voltages can be found by analysis of Fig. 3.4.

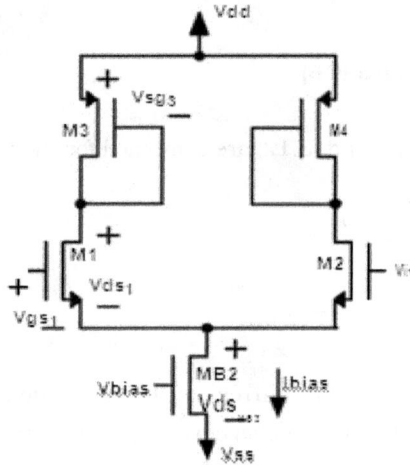

Fig. 3.4 Input Differential Pair with Diode Connected Load

The minimum input voltage can be expressed as

$$V_{IN}^{MIN} = V_{SS} + V_{DSMB}^{SAT}{}_2 + V_{GS}^{Q}{}_1 = V_{SS} + V_{DSMB}^{SAT}{}_2 + V_{DS}^{SAT}{}_1 + V_{THN1} \qquad (3.16)$$

50

and substituting

$$V_{DS}^{SAT} = \sqrt{\frac{2I_D L}{WKP}} \tag{3.17}$$

The minimum input voltage becomes

$$V_{IN}^{MIN} = V_{SS} + \sqrt{\frac{2I_{BIAS}L_{MB2}}{W_{MB2}KP_N}} + \sqrt{\frac{I_{BIAS}L_1}{W_1 KP_N}} + V_{THN1} \tag{3.18}$$

Where, V_{THN1} is the threshold voltage. The minimum input voltage is inversely proportional to the widths of transistors M1, MB2 and directly proportional to the bias current. To reduce V_{IN}^{MIN} the bias current must be reduced or the widths of the input transistors must be increased.

The maximum input voltage can be expressed

$$V_{IN}^{MAX} = V_{DD} - V_{SG3}^{Q} - V_{DS1}^{SAT} + V_{GS1} = V_{DD} - V_{SG3}^{Q} + V_{TH1} \tag{3.19}$$

And by substituting of (3.17)

$$V_{IN}^{MAX} = V_{DD} - \left[\sqrt{\frac{I_{BIAS}L_3}{W_3 KP_P}} + V_{THP3} \right] + V_{THN1} \tag{3.20}$$

V_{THN1} is body effected and will be larger. In this case, the body effect actually increases input range by contributing to V_{IN}^{MAX}. These results indicate that the bias current must be reduced and the width of M3 must be increased to increase V_{IN}^{MAX}. The maximum input voltage of the circuit is, only limited by a V_{GS} drop across transistor M3. For this reason, the input voltage range is typically limited by V_{IN}^{MIN}. The CMR of the NMOS input differential pair is capable of swinging further in the positive direction than the negative direction.

3.4.2 Output Voltage Range

The output voltage range is defined as V_{OUT}^{MAX}, V_{OUT}^{MIN} which represents the maximum output swing available. The output range of the conventional operational transconductance amplifier is reduced due to cascoding at the output shown in Fig. 3.5.

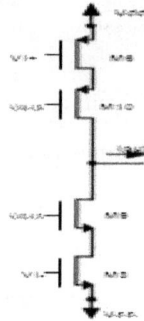

Fig.3.5 Conventional OTA Cascoded Output

The output voltage range is given as

$$V_{OUT}{}^{MAX} = V_{DD} - V_{DS,SAT6} - V_{DS,SAT10} \qquad (3.21)$$

$$V_{OUT}{}^{MIN} = V_{SS} + V_{DS,SAT8} + V_{DS,SAT} \qquad (3.22)$$

3.4.3 Power Dissipation

The static power dissipation (P_{STATIC}) is the product of the sum of the currents flowing through the current sources or sinks with the power supply voltages and is given by

$$P_{STATIC} = (V_{DD} - V_{SS})[I_{D,M1} + I_{D,M2} + I_{D,M5} + I_{D,M6} + I_{D,MB1}] \qquad (3.23)$$

and in terms of I_{BIAS} and K (Fig.3.1)

$$P_{STATIC} = (V_{DD} - V_{SS})I_{BIAS} (2 + K) \qquad (3.24)$$

An increase in the mirror gain factor (K) will increase the slewrate and gain bandwidth of the conventional OTA at the cost of increased area and static power dissipation and a decrease in phase margin.

3.5 Characterization

3.5.1 Gain and Phase Margins

The application of negative feedback requires analysis of the open loop gain. Stability requires a phase shift in the feedback signal less than 180° for open loop gain values larger than 0dB. This requirement necessitates the definition of two measures of stability i.e gain margin (GM) and phase margin (PM). These parameters can be measured via analysis of the open loop AC response simulation. The gain margin is defined as the difference (in dB) in the gain at a phase of -180 degree and unity gain.

Design guidelines typically specify a gain margin greater than 10dB. The phase margin is defined as the difference (in degrees) in the phase at unity gain and -180 degree. The phase margin should be greater than 45° with an optimum, critically damped, value of 60° [5]. For PM values less than 60° the system is under-damped, and the transient response will indicate increased slew rate at the cost of rise and fall peaking. For PM values greater than 60° the system is over-damped, and the transient response will indicate decreased slewrate. Phase margin depends on the relative position of the high frequency pole f_{pA} (Figure 3.3) and the gain bandwidth (unity gain frequency). High frequency pole location is directly related to the phase margin.

3.5.2 Input Offset Voltage

Ideally, if both inputs of the OTA are grounded, the output voltage should be zero [4]. Offset will be present due to random and systematic errors.

Random errors are occurring due to mismatches in the input stage as a result of fabrication, threshold voltage differences and geometric differences of the device. These can be estimated via Monte Carlo simulations.

Systematic errors are due to the circuit design. It can be the result of non-symmetries in the OTA design, creating voltage and current in devices mismatches. It can be determined via simulation and will be evident in the DC sweep simulation as the offset from the zero-zero intercept where the input voltage and output voltage should both equal zero.

3.5.3. Total Harmonic Distortion (THD)

Where the desired output is the fundamental $a_1 V_M \sin(\omega t)$ and ideally, a_2 through a_n are zero [3]. The nth term harmonic distortion is then given by

$$HD_n = \frac{a_n}{a_1}, \quad n > 1 \qquad (3.25)$$

And the total harmonic distortion of the amplifier is given by

$$THD = \sqrt{\frac{a_2^2 + a_3^2 + a_4^2 + \cdots + a_n^2}{a_1^2}} \qquad (3.26)$$

The THD provides a measure of the ratio of the magnitude of output signal harmonics to the expected fundamental output.

3.5.4 Noise

The dominant sources of noise in metal oxide semiconductor transistors are thermal and flicker noise. Fig. 3.6 shows a schematic of a transistor with both thermal and flicker noise sources.

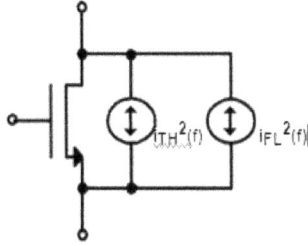

Fig.3.6 MOSFET With Thermal ($i_{TH}^2(f)$) and Flicker ($i_{FL}^2(f)$) Noise Sources.

Thermal noise is causes due to thermal excitation of charge carriers in a conductor [5] and it is proportional to temperature. Thermal noise can be modeled as a current source $i_{TH}^2(f)$ (Figure 3.6) in parallel with the transistor. The thermal noise (current source) can be approximately modeled with the relation $i_n^2(T) = 4KT\left(\frac{2}{3}gm\right)$ where k is Boltzmann's constant ($1.38*10^{-23}$ JK^{-1}), T is the temperature in Kelvin, gm is the transconductance of the transistor, and Δf is the bandwidth in hertz [6]. Flicker noise, it is present under DC conditions and is the result of electron trapping due to silicon imperfections in the transistor. Flicker noise is inversely proportional to the frequency and is commonly referred to as $1/f$ noise. The flicker noise can be modeled with a current source $i_{FL}^2(f)$ in parallel with the transistor as the following equation.

$$i^2(f) = K\frac{I^\alpha}{f}\Delta f \qquad (3.27)$$

Where K indicates the flicker noise constant, Drain current (I_D) is the drain/bias current, α is a constant ($0.5 > \alpha < 2$), and f is the frequency. A simplified model for transistor noise is shown in Fig.3.7.

55

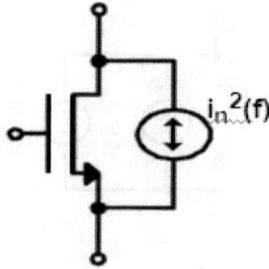

Fig.3.7 Simplified MOSFET Noise Model

$$i_n^2(f, T) = 4KT\left(\frac{2}{3}gm\right)\Delta f + K\frac{I_D^a}{f}\Delta f \qquad (3.28)$$

Above equation is equivalent noise source expression. It is a combination of the thermal and flicker noise source equations. This model is accurate for long channel devices (>1μm). For short channel devices, the thermal noise may be 2 to 5 times larger than shown. The MOS transistor equivalent noise current source (i_n^2), it can also be represented as a voltage source (v_n^2) in series with the gate of the transistor based on the relation $v_n^2 = i_n^2/gm^2$.

The noise contribution of the output stage cascoding transistors (M9,10) shown in   Fig.3.8 is negligible. Cascode transistor noise sources, modeled as voltage sources $v^2_{n,M9}$ and $v^2_{n,M10}$, introduce a small voltage differential at the gate of transistors M9, M10 respectively. This small voltage differential is coupled to nodes X and Y (drains of M6, M9) through M9, M10. Transistors M6, M9 are operating in saturation and their drain current is therefore insensitive to small changes in their V_{DS} voltages. Thus, cascode transistors don't contribute noise to the circuit.

56

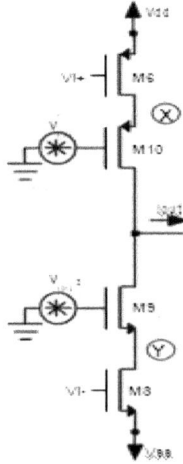

Fig.3.8 Cascode Output Stage With Cascode Transistor Noise Sources

$(V_n{}^2)$

Fig.3.9 shows the one stage conventional operational transconductance amplifier structure with MOSFET noise sources. Assuming the matching transistors (M1=M2, KM3=KM4, M5=M6, M7=M8), neglecting the noise introduced by cascoding transistors M9, M10, and neglecting common mode noise introduced by the biasing transistors, the input referred noise can be derived in terms of flicker noise constant (K) and equivalent noise current source $(i_n{}^2)$ as follows.

$$i_{m,OTA}^2 = 2\left[i_{n.M1}^2 + i_{n.M3}^2 + \frac{1}{K^2}(i_{n.M5}^2 + i_{n.M7}^2)\right] \qquad (3.29)$$

For unity mirror gain (K=1) the equivalent noise is just the summation of transistors M1-M8 noise sources. The noise in transistors M5-M8 decreases with a factor $1/K^2$, indicating a decrease in noise with an increase in current.

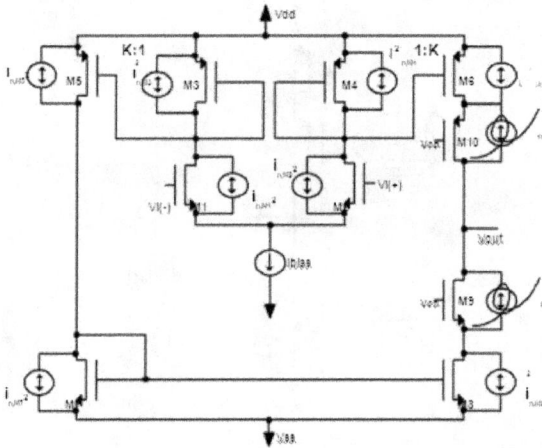

Fig.3.9 Conventional OTA with MOSFET Noise Sources

3.6 Fully Differential (FD) Implementation

3.6.1 Structure

A conventional, fully differential, implementation of the one stage OTA with common mode feedback circuitry is shown in   Figure 3.10. Fully differential structure advantages are improved output voltage swing, less susceptibility to common mode noise, and cancellation of even-order nonlinearities.

58

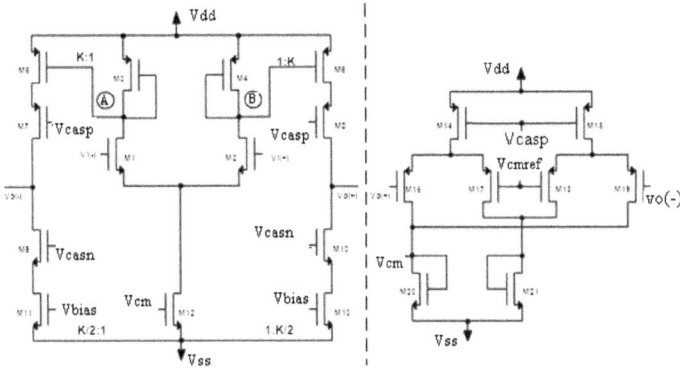

Fig.3.10 FD Conventional OTA and Common Mode Feedback Circuit

The architecture of the fully differential OTA is similar to that of its single ended counterpart (Figure 3.1) with the following exceptions. First, the mirror in the single ended architecture, formed by transistors M7 to M8, has been replaced with cascoded current mirrors, biased by V_{biasn}, and formed by M9, M11 / M10, M13. Second, a common mode feedback circuit formed by transistors M14 to M21 has been implemented to control the common mode output voltage. The bias current of the differential pair is now controlled by the common mode feedback circuit via the control voltage V_{cm}. This design lacks current mirroring capability generated by transistors M7, M8 for the single ended structure and is therefore only capable of a maximum output current at each output equal to half the circuit bias current. In order to achieve output currents equal to the bias current at each output, the dual shell structure shown in Fig.3.11 was implemented.

This fully differential OTA maintains mirroring capability via mirrors formed by transistors M7, M8 and transistors M13, M14 and is capable of generating maximum output currents equal to I_{BIAS} at each output terminal.

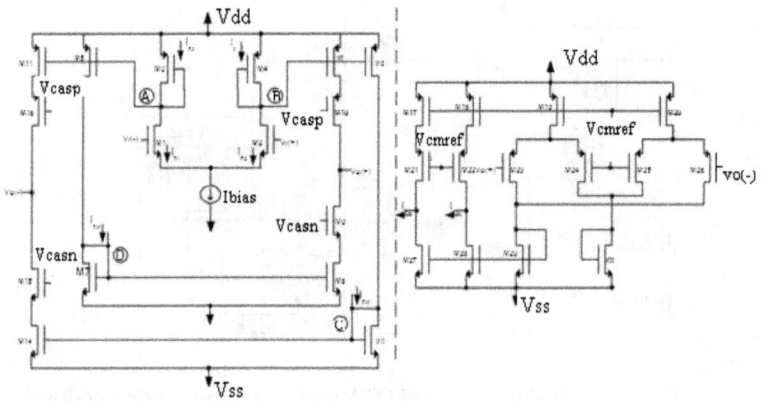

Fig. 3.11 Implemented Dual Shell Fully Differential Conventional OTA

3.6.2 Common Mode Feedback

The implemented Common Mode Feed Back (CMFB) circuit for the dual shell structure (Figure 3.11) is more complex than that of the common mode feedback circuit for the standard fully differential OTA (Figure 3.10) but still utilizes dual input differential pairs and performs the same function. The dual differential pair structure introduces only a small capacitive load ($C_{gs16, 19}$) at the output of the operational transconductance amplifier avoiding resistive loading common to many common mode feedback circuits. The function of the common mode feedback circuit is to define/control the common mode voltage such that

$$V_0(+) = V_0(-) = V_{CMref} = V_{CM} \qquad (3.30)$$

when a common mode input is applied ($V_0(+) = V_0(-)$).

V_{CMref} is typically set to

$$V_{CMref} = \frac{V_{DD} - V_{SS}}{2} = V_{CM} \qquad (3.31)$$

such that for complementary sources, $v_{cm}=0$. The common mode reference voltage (V_{cmref}, Figure 3.10) is compared with $v_{o,cm}$ is given by

$$V_{O,CM} = \frac{V_0(+) - V_0(-)}{2} \qquad\qquad (3.32)$$

The common mode voltage of the conventional, fully differential, dual shell, OTA is controlled via current injection from the common mode feedback circuit at low impedance nodes D/E (Fig. 3.11).

The CMFB circuit utilizes two, identical, current mirrors, formed by transistors M27, M29 and M28, M29 to generate replica correction currents (I_{CM}) which are injected at low impedance nodes D/E. These current mirrors are needed to provide correction currents for each complementary shell independently. Active control of the circuit bias current will not correct the common mode voltage of the dual shell FD OTA due to the mirroring function of the shell structures.

Adjustment of the bias current would result in complementary voltage changes at nodes A/B, resulting in complementary drain current changes for transistors M5, M6 and transistors M11, M12. The shell type structure would mirror these complementary changes to the circuit output, resulting in zero net voltage change. Replica common mode correction currents are therefore required to facilitate independent shell common mode correction. Based on this analysis, two common mode voltage corrections are possible (referring to fig. 3.11) as follows:

1. If $v_{o,cm} > v_{cmref}$, less current is passed through M29. This decrease in current is mirrored to M27, M28. Currents $I_{D27,28}$ are fixed by transistors M17, M18 respectively. The excess current $i_{d17,18}$ generated by these transistors is then injected at nodes E/D (transistors M7, M13) and coupled to the output via mirror transistors M8, M14. The resulting increase in $i_{D8,14}$ pulls current from the output node and the output voltage is reduced, to attain $v_o(+) = v_o(-) = v_{cmref}$.

61

2. If $v_{o,cm} < v_{cmref}$, more current is passed through M29. This increase in current is mirrored to M27, M28. $I_{D27,\ 28}$ is fixed by transistors M17, M18 respectively. The required increase in current is then pulled from nodes D/E (transistors M7, 13) and coupled to the output via mirror transistors M8, M14. The resulting decrease in $i_{D8,14}$ injects current at the output node and the output voltage is increased, to attain $v_o(+) = v_o(-) = v_{cmref}$.

3.6.3 Stability

Since the CMFB loop is a negative feedback loop, stability is a key issue [4]. The CMFB structure has a dominant low frequency pole (f_{pIN}) at the input (due to C_L) and four high frequency poles associated with diode connected transistors M27, M28 and the feedback injection transistors M7, M13 (nodes D/E, fig. 3.11).

3.7 Conclusion

A full analysis of the one-stage conventional OTA structure is presented. The transconductance amplifier generates an output current dependent on the circuit input voltage and rejects applied common mode signals. Cascoding of the output shell structure increase the gain necessitates the requirement of a purely capacitive load. The structure is classA and the maximum output current is limited by the magnitude of the circuit bias current in the output shell. The mirror gain factor can be increased to increase the maximum output current (slewrate) and gain bandwidth at the cost of an increase in static power and area. The mirror gain factor decrease in phase margin and stability.

CHAPTER 4

GAIN BOOSTED OPERATIONAL TRANSCONDUCTANCE AMPLIFIER

4.1 Introduction

Significant and effectiveness Analog to Digital converters and switched capacitor filters need operational transconductance amplifier which consist large DC gain and a considerable extent Gain Band Width product (GBW). The approach of deep submicron technologies enables progressively large speed circuits, but generates designing high DC gain operational transconductance amplifier more challenging. Analog device and mixed signal systems parameters like speed and accuracy are confined by the settling time behavior of the CMOS operational transconductance amplifiers. Dynamic biasing of amplifiers has been recommended as a method for conventional techniques, suggesting that fast settling need one pole settling behavior and large gain bandwidth product.

Adaptive biasing circuits design techniques improves speed and gain bandwidth product which in turn complicate the design and reduce circuit input CM voltage range. The design techniques proposed in this thesis solve the issues mentioned above, and combines excellent performance with simplicity of design, which is suitable for high frequency applications. With these design techniques settling time increases and also improves small signal characteristics like Gain Band Width (GBW) product, DC gain, and slewrate without consuming extra power.

Speed and accuracy are the two most significant properties of analog circuits. Optimize architecture for both direction shows to conflicting demands. In large applications of CMOS analog circuits such as pipeline analog to digital converters and switched capacitor filters, accuracy and speed are calculated by the settling behavior of opamps. Fast settling needs large unity gain frequency and a one pole settling behavior of the operational amplifier and accurate settling needs a large dc gain. The realization of a CMOS opamp that requires large dc gain with high unity gain frequency has been a difficult problem.

Requirement of high gain need multistage architecture designs with long channel devices biased at low current levels, whereas the high unity gain frequency requirement calls for a single stage design with short channel devices biased at high circuit bias current levels. There have been several architecture approaches to circumvent this problem. This section describes OTA with LCMFB and the design of the OTA used in the 10-bit, 100MHz pipelined analog to digital converter. First the basic theory of gain boosting and the high frequency behavior are explained and then the design of gain boosted telescopic OTA is explained clearly step by step.

4.2 Design of Low voltage and high gain OTA specifications

Consider specific application to design low voltage and high gain single stage operational transconductance amplifier. The Analog to Digital converter application has been considered with specifications of 10 bit resolution, 100MHz of sampling rate, input voltage range of 1Vp-p with latency less than 5 clock cycles. Based on the specifications pipeline architecture with 2.5 bit/stage is chosen. This requires a residue amplifier of gain 4.

OTA requires the following parameters to design 10 bit A/D converter with 2.5V power supply.

4.2.1 Open loop DC gain

The closed loop gain of an amplifier with a feedback factor of β is given by the equation

$$A_f = \frac{A}{1+A\beta} \qquad (4.1)$$

Where A_f is the closed loop gain, A is the open loop gain and β is the feedback factor.

So the required ideal value for A_f is $\frac{1}{\beta}=4$.

Now the error due to finite open loop DC gain (A) is given by

$$\Delta = \frac{1}{\beta} - \frac{A}{1+A\beta} \qquad (4.2)$$

This error must be less than 1/2LSB.

i.e.,

$$\Delta < \frac{1}{2}LSB < \frac{1}{2^{11}} \qquad (4.3)$$

Solving this we will get $A > 2^{15} - 4$

Thus the calculated open loop DC gain is nearly equal to 110db.

4.2.2 Unity Gain Frequency (UGF)

For a switched capacitor amplifier in hold mode UGF or GBW is given by

$$GBW > N.\ln2.\frac{C_{L,tot}}{2\pi T \beta C_L} \qquad (4.4)$$

Where N-resolution of ADC=10, β- Feedback factor=0.25

C_L-Load Capacitance=1pF, $C_{L,tot}$- Effective load capacitance=1.3pF

T- Settling time is generally 2/3rd of the time period i.e., $T = \frac{2}{3} \times \frac{T_s}{2}$

Where $T_s = \frac{1}{f_s}$

Therefore

$$GBW > \frac{10*ln2*1.3p}{2*\pi*\left(\frac{1}{3}*100M\right)*0.25*1p} = 1.72GHz.$$

4.2.3 Output Swing

Since the input and output of each stage of the pipelined ADC has to be within 1 Vp-p, the output swing of the OTA is also 1 Vp-p.

4.2.4 Phase Margin

The phase margin must be greater than 60 degrees, which is needed for fast settling and stability. We required Open loop DC gain of 110db, 1.72GHz of gain bandwidth, Output swing of 1Vp-p with phase margin greater 60 degree.

4.3 Proposed OTA with LCMFB

The proposed operational transconductance amplifier with local common mode feedback is shown in Fig.4.1. Similar to the conventional operational transconductance amplifier, the local common mode feedback OTA architecture consist a differential pair M1 and M2. This architecture is having three current mirrors M3 M5, M7, M8, and M4, M6. In the local common mode circuit the active load transistors M3, M4 are reconnected to have a common gate i.e node C and matched resistors R1 and R2 are used to connect the gate and drain terminals of M3, M4. This simple modification has several performance enhancing benefits versus the conventional operational transconductance amplifier architecture including class AB operation which provides enhancement in slewrate, gain bandwidth, and linearity, with equal static power dissipation.

Implementation of resistors R1, R2 with MOS transistors MR1, MR2, shown in Figure 4.2, provides programmable performance characteristics (via the control voltage V_R), allowing utilization of the same operational transconductance amplifier for several applications. Class AB operation characteristics allow the local common mode feedback structure to outperform the conventional structure with unity current mirror gain. The analysis for the local common mode feedback OTA will therefore be based on a unity mirror gain factor (K=1, transistors M3=M4=M5=M6 and M7=M8).

66

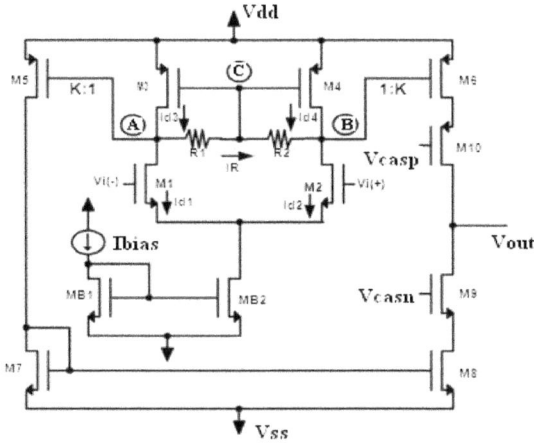

Fig.4.1 OTA with Local Common Mode Feedback

4.3.1 Operation

Fig.4.2 shows the local common mode feedback OTA structure with transistors MR1, MR2 implemented to function in the triode region and act as programmable resistors.

For quiescent or common mode operation, the drain currents of transistors M1 to M10 have equal values ($I_{D1-10}=I_{bias}/2$) while the current i_R in transistors MR1, MR2 is zero. The gate source voltage of transistors M3, M4 is the same as their drain-source voltage. For common mode input signals, transistors perform as less impedance loads with value.

$$R_L^{CM} = \frac{1}{gm_{3,4}}$$ (4.5)

67

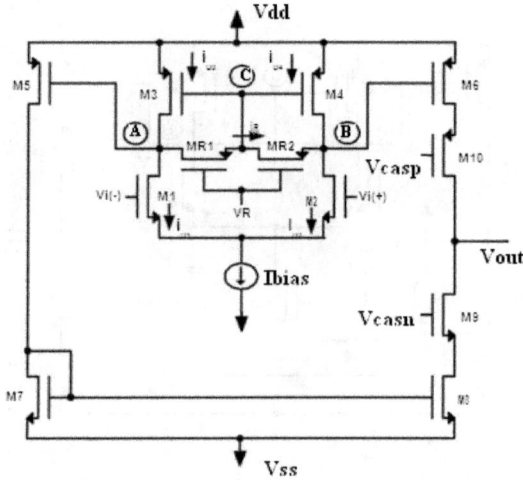

Fig.4.2 LCMFB OTA with MOS Transistors MR1, MR2

Upon application of a differential signal, the signal current component (i_d=i_r) lows through transistors MR1, MR2, and $i_{D1,2}$ are given by

$$i_{D1,2} = I_D + i_d \qquad (4.6)$$

Where,

$$i_d = gm_{1,2}\frac{v_d}{2}\sqrt{1 - \frac{\frac{v_d}{2}}{V_{GS1,2}-V_{THN1,2}}} \qquad (4.7)$$

and v_d is the applied differential voltage. The drain currents in M3, M4 is not changed ($i_{D3,4}$=I_{BIAS}/2). The current i_r generates differential complementary voltage changes between nodes A and B with C node remains at a signal constant voltage. Various nodes A and B voltages are given by

$$V_A = -V_B = i_r R_{MR1,2} \qquad (4.8)$$

Where $R_{MR1,2}$ is taken as the resistance generated by transistors MR1, MR2 and, based on the triode channel resistance. Equation is given by

$$R_{MR1,2} = \frac{1}{\beta_{MR1,2}(V_C-V_R-V_{THP})} \qquad (4.9)$$

Where V_R is the applied control voltage (Figure 4.2), $\beta_{MR1,2}$=$KP(W_{MR1,2}/L_{MR1,2})$, and V_C is the constant voltage at node C. This complementary swing at node A and B produces high, non-complimentary, signal currents in the shell (M5 to M10) of the operational transconductance amplifier by generating high gate-source voltage differentials for common source transistors M5, M6, respectively.

4.3.2 Signal Analysis

Determination of the open loop gain of the local common mode feedback operational transconductance amplifier circuit needs two independent analyses. The differential input of the circuit must first be calculated followed by the common source circuit output shell. The collective local common mode feedback operational transconductance amplifier gain will then be defined as the combination of the past analyses. The following analysis will assume transistor matching as $M_1=M_2$, $M_3=M_4=M_5=M_6$, $M_7=M_8$, and $MR_1=MR_2$. The gain will be analyzed by negative input terminal (at the gate of M_1) to the output terminal. Analysis of the circuit positive signal input (at the gate of M_2) would give similar results with equivalent transistor substitutions as listed above.

The differential input stage is shown in Fig.4.3.

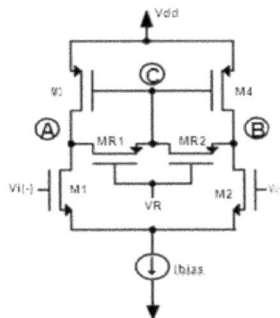

Fig.4.3 LCMFB OTA Differential Input Stage

Presented that the common combined gate of transistors M3, M4 (node C) is an AC ground; the architecture can be simplified to the analysis circuit and small signal models (derived from the MOSFET hybrid-π model) as shown in Fig.4.4 (a), (b), and (c).

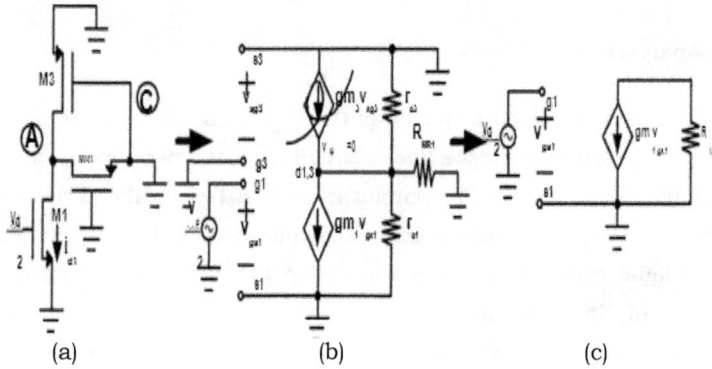

(a)   (b)   (c)

Fig.4.4 (a) Input Stage Simplified Gain Analysis Schematic (b) Small Signal Equivalent Circuit (c) Simplified Small Signal Equivalent Circuit

The equivalent resistance is given by

$$R_{A,B} = r_{o12} \parallel r_{o3,4} \parallel R_{MR1,2} \tag{4.10}$$

$v_{A,B}$ are then a combination of (4.6) and (4.7) and is given by

$$V_{A,B} = i_{d1,2}R_{OUT} = (r_{o12} \parallel r_{o3,4} \parallel R_{MR1,2})(gm_{1,2}v_{gs1,2}) \tag{4.11}$$

$v_d/2 = v_{gs1,2}$ and the gain of the differential (A_{CORE}) input stage is given as

$$A_{CORE} = \frac{V_{A,B}}{v_d} = \frac{gm_{1,2}R_{A,B}}{2} \tag{4.12}$$

The common source output shell is shown in Figure 4.5 and consists of two common source amplifiers M_5 and M_6, a current mirror M_7 and M_8, and cascoding output transistors M_9 and M_{10}. The cascoded output stage provides the large output resistance which is required for large gain in the output shell.

70

The output resistance of the local common mode feedback OTA is similar to that of the conventional operational transconductance amplifier and is given by

$$R_{OUT} = gm_{10}r_{o10}r_{o6} // gm_9r_{o9}r_{o8} \qquad (4.13)$$

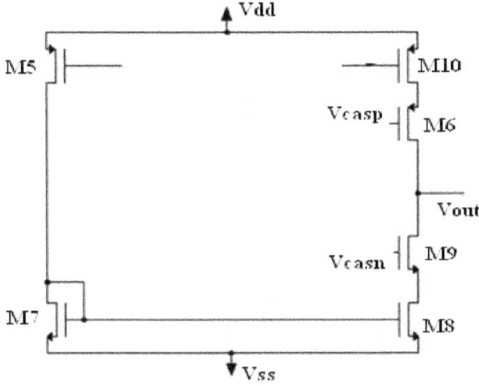

Fig.4.5 LCMFB OTA Common Source Output Shell

The current mirror has unity gain, and the gain of the output stage (A_{OS}) is given by

$$A_{SHELL} = 2gm_{5,6} \ R_{OUT} = 2gm_{5,6} \ (gm_{10}r_{o10}r_{o6} // gm_9r_{o9}r_{o8}) \qquad (4.14)$$

Local common mode feedback OTA open loop gain is given by the multiplication of the gain of the circuit input stage (A_{CORE}) and the output shell (A_{SHELL}) and is given by

$$A_{OL} = gm_{1,2} \ R_{A,B} \ gm_{5,6} \ R_{OUT} \qquad (4.15)$$

This definition indicates that the gain of the local common mode feedback OTA is a function of the programmable resistance R_{MR1} and R_{MR2}. For $R_{MR1,2} \approx 1/gm$, $R_{MR1,2}$ will dominate the parallel combination $r_{o1,2}//r_{o3,4}//R_{MR1,2}$,

71

and the structure will behave as a one stage amplifier ($A_{CORE} << A_{SHELL}$). An increase in $R_{MR1,2} \approx r_{o1,2} // r_{o3,4}$ will result in increased gain in the input stage and the structure will behave as a two stage amplifier ($A_{CORE} < A_{SHELL}$). Two stage operations would eliminate the need for cascoding transistors M9, M10 and would require compensation.

4.3.3 AC Analysis

Fig.4.6 is referred for AC analysis. Similar to the conventional OTA, the frequency response of the local common mode feedback operational transconductance amplifier is calculated mainly by low impedance, large frequency, poles at nodes A/B, in addition with the large impedance and low frequency pole at the circuit output node. The following analysis defines the high frequency pole and assumes nodes A and B as equivalent nodes in terms of resistance and parasitic capacitance (M1=M2, M3=M4, M5=M6, MR1=MR2). The resistance at nodes A/B becomes a function of the triode resistance ($R_{MR1,2}$) created by MR1, MR2 and is given by

$$R_{A,B} = r_{o1,2} \parallel r_{o3,4} \parallel R_{MR1,2} \qquad (4.16)$$

Given that node C (Figure 4.6) is a virtual ground, the parasitic capacitance at nodes A/B does not include $C_{gs3,4}$. The addition of MR1, MR2 does introduce an additional parasitic $C_{sb,MR1,2}$, but the well known relation $C_{gs} >> C_{sb}$, and the relative dimensions of the transistors $W_3 >> W_{MR}$, indicate the parasitic capacitance is decreased by a factor close to 2 ($K=1$) versus the single conventional OTA structure. The parasitic capacitance at A/B is given by

$$C_{A,B} \approx C_{gs5,6} \qquad (4.17)$$

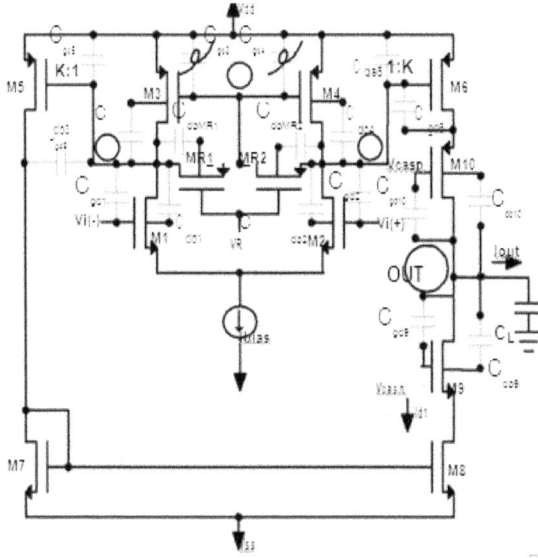

Fig.4.6 LCMFB One Stage OTA AC Analysis Schematic

Combining equations (4.16) and (4.17), the pole at A/B is

$$f_{pA,B} = \frac{1}{2\pi C_{A,B} R_{A,B}} = \frac{1}{2\pi C_{gs5,6}(r_{o1,2} \| r_{o3,4} \| R_{MR1,2})} \tag{4.18}$$

The capacitance of output node is controlled by the load capacitance and is similar to the single conventional OTA architecture.

$$C_{OUT} = C_{dg9} + C_{db9} + C_{dg10} + C_{db10} + C_L \approx C_L \tag{4.19}$$

Combining equations (4.18) and (4.19), the pole at A/B is

$$f_{pOUT} = \frac{1}{2\pi C_{OUT} R_{OUT}} = \frac{1}{2\pi C_L (gm10ro10ro6 \,//\, gm9ro9ro8)} = f_{3db} \tag{4.20}$$

Gain Bandwidth product is

$$GB = \frac{[gm_{1,2}gm_{5,6}(r_{o1,2}//r_{o3,4}//R_{MR1,2})]}{2\pi C_L} \qquad (4.21)$$

The GB is dependent on the variable resistance $R_{MR1,2}$. As $R_{MR1,2}$ increases, the gain bandwidth increases.

4.3.4 Maximum Output Current

The maximum output current can be found by analysis of Fig.4.7. Class AB operation provides large non-symmetric currents in the circuit output shell. These currents are generated by the large gate-source voltage swings (generated at nodes A/B) applied to transistor M5, M6. Maximum output current produces when the maximum gate and source differential is applied to transistor M6 (or M5) and M5 (or M6) is in cutoff.

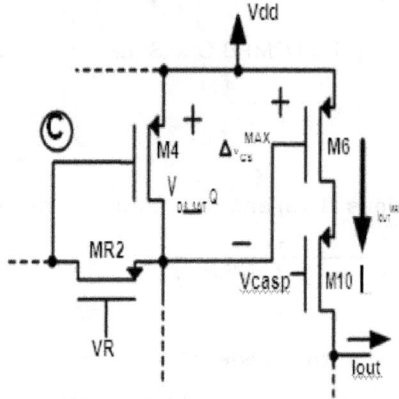

Fig.4.7 LCMFB OTA Maximum Output Current Schematic

The maximum output current is given by

$$I_{OUT}^{MAX} = \beta_{5,6}(V_{SD,SAT3,4}^Q + \Delta V_{GS5,6}^{MAX})^{\wedge}2 \qquad (4.22)$$

Where,

74

$$\Delta V_{GS5,6}^{MAX} = \frac{I_{OUT}^{MAX}}{2} R_{MR1,2} \tag{4.23}$$

represents the maximum swing at nodes A and B.

4.3.5 Slew Rate

The slew rate is then given by

$$SR = \frac{I_{OUT}^{MAX}}{C_L} \tag{4.24}$$

The Slewrate is defined as output current divided by load capacitance.

4.3.6 DC Analysis

The Input CM range of the LCMFB OTA is equivalent to the CMR of the conventional structure. The output voltage range of the LCMFB operational transconductance amplifier is equal to the output range of the conventional structure

The static power dissipation (P_{STATIC}) is the product of the sum of the currents flowing through the current sources or sinks with the power supply voltages and is given by

$$P_{STATIC} = (V_{DD} - V_{SS})\left[I_{D,M1} + I_{D,M2} + I_{D,M6} + I_{D,MB1}\right] \tag{4.25}$$

And in terms of I_{BIAS} (K=1 for the LCMFB structure)

$$P_{STATIC} = (V_{DD} - V_{SS})3I_{BIAS} \tag{4.26}$$

Class AB operation in the local common mode feedback OTA generates signal currents much higher than the bias current applied with the same static power dissipation as that of the single conventional structure (K=1). The advantage of this operation is the capability to design high slewrate architectures with low static power dissipation.

4.3.7 Characterization

The phase margin is an indication of the relative position of the high frequency pole $f_{pA,B}$ and the gain bandwidth. For a fixed gain bandwidth, an increase in the frequency of $f_{pA,B}$ make to an increase in the phase margin. In the capacitance at nodes A/B is translating to an increase in the position of the high frequency pole and an increase in the phase margin. The position of $f_{pA,B}$ is a function of the resistance generated by transistors MR1, MR2 (Figure 4.2).

The control voltage V_R provides programmable resistance values for transistors MR1, MR2 and therefore allows programmable control of the phase margin. Both the phase margin (via $f_{pA,B}$) and the open loop gain are a function of the triode resistance generated by MR1, MR2, $R_{MR1,2}$. A decrease in phase margin increases slewrate at the cost of decreased stability. The resistance $R_{MR1,2}$ can, therefore, be used to trade off slewrate and gain bandwidth enhancement with phase margin.

The definition of the offset voltage for the single stage conventional OTA is equivalent to the local common mode feedback structure. The definition of the total harmonic distortion for the conventional OTA is equivalent to the LCMFB structure.

4.3.8 Noise

The local common mode feedback OTA structure with MOSFET noise sources is shown in Fig.4.8.

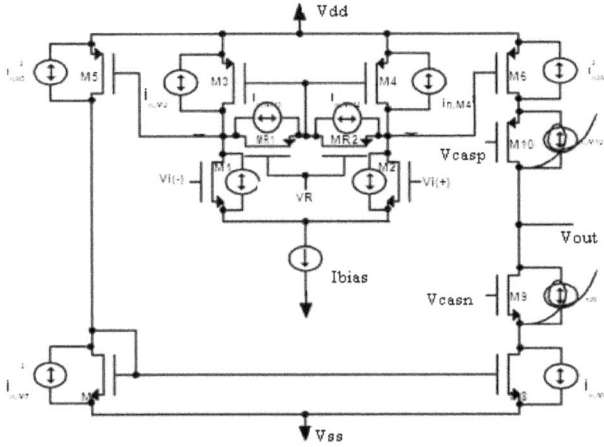

Fig.4.8 LCMFB OTA with MOSFET Noise Sources

Assuming matching transistors (M_1=M_2, M_3=M_4=M_5=M_6, M_7=M_8, MR_1=MR_2), not considering the noise introduced by cascoding transistors M9, M10, and not considering common mode noise introduced by the biasing transistors, the input referred noise can be derived (in terms of i_n^2) as follows.

$$i_{niLCFBOTA}^2 = 2\left[i_{n,M1}^2 + i_{n,M3}^2 + i_{n,MR1}^2 + \frac{1}{(gm_{5,6}R_{A,B})^2}\left(i_{n,M5}^2 + i_{n,M7}^2\right)\right] \quad (4.27)$$

The noise of the LCMFB OTA is a function of the transconductance of M5, M6, and the resistance $R_{A,B}$.

Resistors R_A and R_B are dependent on resistor R_{MR} indicating that an increase in resistor R_{MR} results in a decrease in noise contributed from the structure shell transistors. The addition of transistors MR1, MR2 results in the addition of two sources of noise for the LCMFB configuration ($2*i_{n,MR1}^2$) which contribute negligible additional noise and decrease the noise contribution of the shell transistors.

4.4 Fully Differential Implementation

4.4.1 Structure

A fully differential implementation of the operational transconductance amplifier with local CM feedback and a common mode feedback network is shown in Figure 4.9. The basic structure of the fully differential local common mode feedback architecture is similar to that of its single ended counterpart with the following exceptions. A second shell, formed by transistors M11 to M16, generates complementary output current. Implementation of the second shell structure is required to provide mirroring of large current, class AB currents at the outputs of positive and negative. A common mode feedback circuit, formed by transistors M17 to M30, has been implemented to control the common mode output voltage. This voltage is controlled via current which injection from the common mode feedback circuit at low impedance nodes D/E.

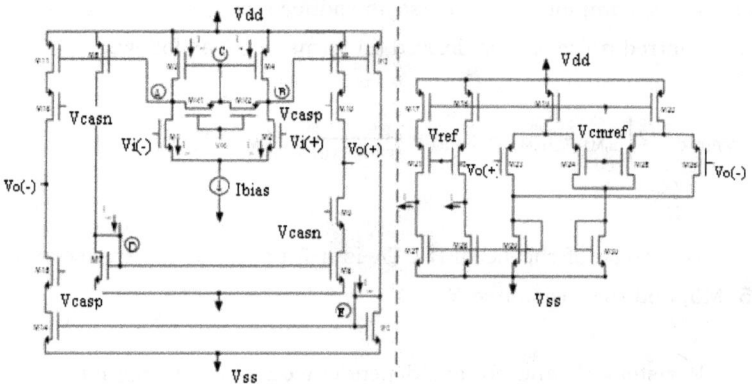

Fig.4.9 Fully Differential LCMFB OTA and Common Mode Feedback Circuit

Advantages of the FD structure include improved voltage swing, less susceptibility to CM noise, and even-order nonlinearity cancellation.

78

4.4.2 Common Mode Feedback

The CM feedback architecture for the local common mode feedback OTA utilizes two identical current mirrors, formed by transistors M27, M29 and M28, M29, to generate replica correction currents (I_{CM}) which are injected at low impedance nodes D/E (Figure 4.9). These current mirrors are required to provide correction currents for each complementary structure shell independently. Active control of the bias current will not correct the common mode voltage of the local CM feedback fully differential OTA due to the mirroring function of the shell structures. Adjustment of the bias current would result in complementary voltage changes at nodes A/B, resulting in complementary drain current changes for transistors M5, M6 and M11, M12. The shell structure would mirror these complementary changes to the circuit output, resulting in zero net voltage change. Replica CM correction currents I_{CM} are therefore need to facilitate independent shell common mode correction. Based on this analysis, and similar to the conventional structure, two common mode voltage correction scenarios are possible (referencing Figure 4.9) as follows:

1. If $v_{o,cm} > v_{cmref}$, less current is passed through M29. This decrease in current is mirrored to M27, M28. $I_{D27,28}$ is fixed by transistors M17, M18 respectively. The excess current $i_{d17,18}$ generated by these transistors is then injected at nodes D/E (transistors M7, M13) and coupled to the output via mirror transistors M8, M14. The resulting increase in $i_{D8,14}$ pulls current from the output node and the output voltage is reduced, to attain: $v_o(+) = v_o(-) = v_{cmref}$.

2. If $v_{o,cm} < v_{cmref}$, more current is passed through M27. This increase in current is mirrored to M27, M28. $I_{D27,28}$ is fixed by transistors M17, M18 respectively. The required increase in current is then pulled from nodes D/E (transistors M7, M13) and coupled to the output via mirror transistors M8, M14. The resulting decrease in $i_{D8,14}$ injects current at

the output node and the output voltage is increased, to attain: $v_o(+) = v_o(-)$ $= v_{cmref}$.

4.4.3 Stability

Stability of the common mode feedback loop is equally important to that of the differential loop. The LCMFB OTA CMFB structure has a dominant low frequency pole (f_{pIN}) at the input (due to C_L) and four high frequency poles associated with diode connected transistors M27, M28 and the feedback injection transistors M7, M13 (nodes D/E, Figure 4.9). The gain and phase margin of the common mode feedback circuit must be verified.

4.5 Gain boosted OTA

A gain boosted OTA was chosen as an optimum choice for our application because of the following points:

- It is easy to design over the other designs and large speed operation because of one-pole.
- Higher power efficiency and consists lower noise factor compared to other OTA designs.
- Its limited output swing is not a constraint for our application as the desired swing is 1 Vp-p, which could easily be achieved using this topology.
- At no point of time output needs to be shorted with input, which otherwise limits input common mode range of the telescopic amplifiers.
- NMOS input transistors were selected as flicker noise will not limit the performance because of its upper cut off frequency is approximately 100 KHz.
- Gain booster architecture provides the gain without affecting the frequency response of telescopic OTA.

4.5.1 Design of Telescopic OTA

The schematic of the telescopic OTA is shown in Fig. 4.10. This configuration has been arrived at after trying out various architectures for achieving the required specifications. Half of the fig. 4.10 shows the telescopic OTA and other half is CMFB (Common Mode Feed Back) circuit. Input and output pins represent connections to nodes in the circuit. The input transistors (M1 and M5) were sized with a high W/L aspect ratio to enhance the high transconductance need to quickly move charge onto the test capacitors. Transistors M9, M10, M11 and M12 form a cascode load. M0 is used to provide biasing for the OTA. Ensure that all the transistors are must operate in saturation region. If transistors operate in the triode region it will cause the behavior of the operational transconductance amplifier to be non-linear. It will produce poor transient response as well as a less DC gain.

To start the design of operational transconductance amplifier we need to know the bias current that is required to meet the required specifications. It depends on the slewrate.

Slewrate: Slewrate individually determines the performance of the OTA. The output swing needed is 1Vp-p. The clock frequency is 100MHz. Generally 1/3rd of the time is reserved for slewing.

So the time for slewing is given by $t = \frac{1}{3*2*100M}$.

Hence the   slew rate= $\frac{\text{output voltage swing}}{\text{time}} = \frac{1v}{1.66ns} = 0.6\frac{v}{ns}$.   (4.28)

The bias current through transistor M0 is,

   I= slew rate * load capacitance= 1.2mA.

If current I increases further then slewrate increases, which is better but the power dissipation also increases. If the power dissipation is within the given range then current can be increased to get fast settling. The current (I) has been divided equally into the two input transistors M1 and M5. In analog circuit for the design of amplifiers all the transistors must be operated in the saturation region. I-V equation of the MOSFET in saturation region is given

$$\frac{W}{L} = \frac{2I_d}{kV_{od}^2}$$ (4.29)

Where $k = \mu C_{ox}$ and $V_{od} = V_{gs} - V_{th}$.

The Transconductance of the transistor g_m is given by

$$g_m = 2I_d/V_{od}$$ (4.30)

Substituting V_{od} in W/L gives,

$$\frac{W}{L} = \frac{g_m^2}{2kI_D} \text{ where } V_{od} = \frac{2I_d}{g_m}$$ (4.31)

Gain Bandwidth Product (GBW) = $g_m r_o * \frac{1}{r_o C_L}$ (4.32)

$$g_m = \frac{(2\pi f_u)^2}{2kI_D}$$ (4.33)

From this, sizes of input transistors M1 and M5 can be calculated.

The same current flows through the two legs of telescopic OTA. So I_d for the remaining transistors is also the same. Their W/Ls depends on the output voltage swing. Initially we have to assume overdrives for all these transistors in such a way that the output swing specified has been satisfied and the transistors must be in saturation. The bias voltages Vbn and Vbp should be chosen based on the overdrive voltages.

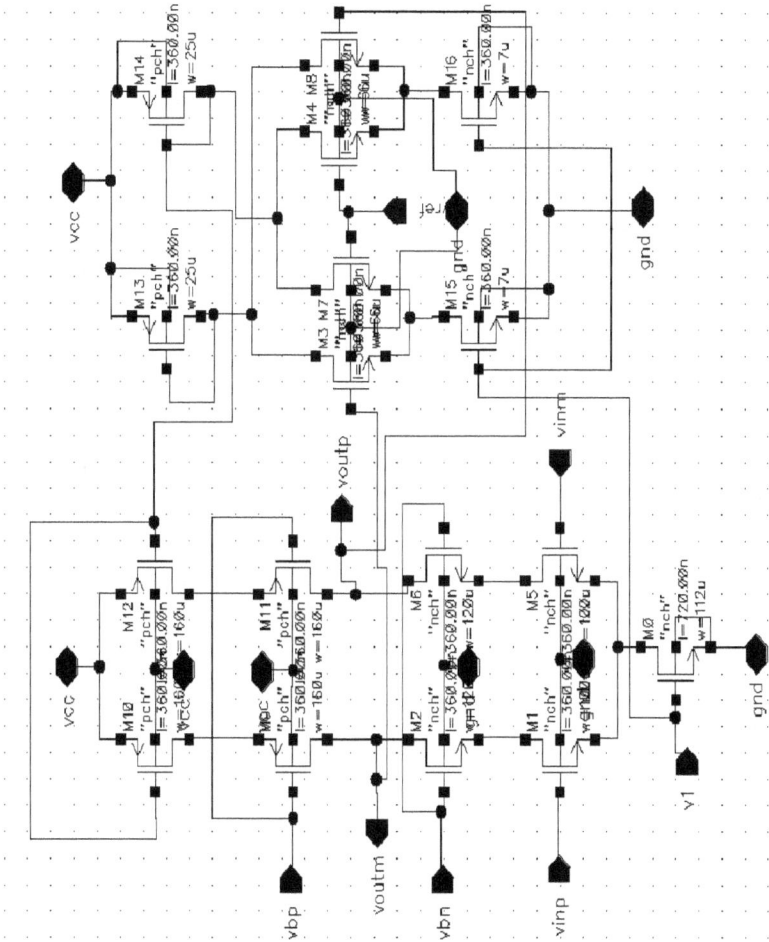

Fig.4.10 Schematic of Telescopic OTA

Since the mobility of p-channel is lower than mobility of n-channel MOSFET, the transconductance parameter k is less for p-channel MOSFET. So in order to keep the size of transistor to a reasonable value for the same drain current the overdrive voltage for p-channel MOSFET must be kept high.

$$\frac{W}{L} = \frac{2I_D}{KV_{od}^2} \qquad (4.34)$$

The values of K_p = 53$\mu A/V^2$ and K_n = 230 $\mu A/V^2$.

4.5.2 Common-Mode Feed Back (CMFB) in Gain Boosted OTA

CMFB is necessary in a fully differential OTA to keep the outputs from drifting high or low out of the range where the amplifier provides high of gain. Continuous time CMFB has been chosen for our OTA design. In fact, several reasons influence to this choice. First, the use of switch engenders non linearity at the output more than with a continuous time structure. Then the use of digital signals in the first stage can generate additional noise and parasitic in general. The non use of digital signal in this stage avoids several coupling problems and layout care that would have to be taken in account. CMFB circuit shown in Figure 4.11 is based on a standard design [2].

It uses two differential pairs (M3, M7 and M4, M8) to sense the difference between the average output voltage and a common mode voltage V_{CM} or V_{ref} (1.65V in this case) which is supplied externally. V_{cmfb} is used to bias a transistor that adds to the bias current and keeps the common mode from drifting. The current in the CMFB circuit need not be large as long as the currents through the top and bottom of the OTA are fairly well balanced. The current through CMFB is small so that the power dissipation is also small.

The details of Widths/Lengths of all the transistors are given in Table.4.1.

Transistor	Size(μ/μ)
M0	112/0.72
M1, M5	80/0.36
M2, M6	120/0.36
M9, M10, M11, M12	160/0.36
M13, M14	25/0.36
M3, M4, M7, M8	6/0.36
M15, M16	7/0.36

Table 4.1 W/Ls of Telescopic OTA

Bias voltages for the OTA are Vbn=2.5V and Vbp=2V.

The telescopic OTA alone meets all the desired specifications except the open loop DC gain. So gain boosted OTA is needed since the response does not change with this frequency. The architecture used in this design consists n-channel input differential pair and the p-channel cascode load. So here, the design of two auxiliary amplifiers used is discussed.

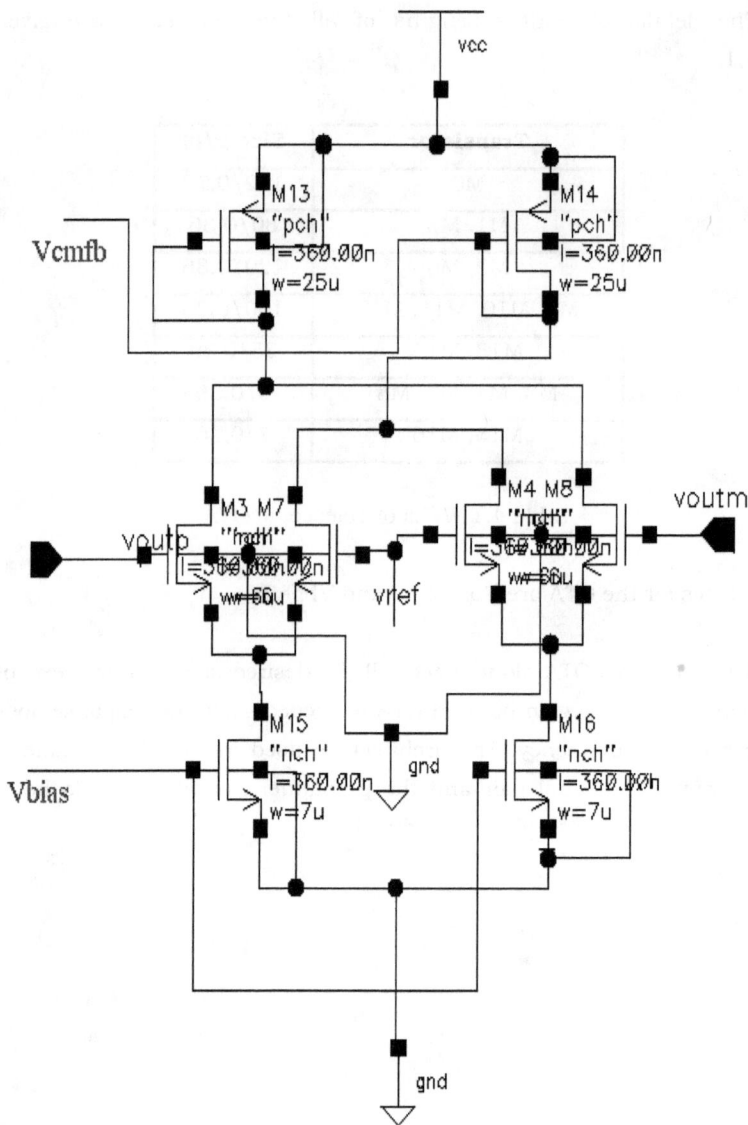

Fig.4.11 CMFB circuit.

4.5.3 Design of Auxiliary Amplifiers

As seen from the previous section the bias voltages for the OTA are 2.5V and 2V. With the auxiliary amplifiers we have to provide that bias voltage. Here the bias voltages are near to the supply voltage. Using simple differential amplifiers we get that bias voltages, but the gain is low. Similarly Telescopic OTA has the limited swing, so we are not able to get the required bias voltages using the telescopic architecture. Hence in this design folded cascode OTAs are used as the auxiliary amplifiers that can provide the mentioned bias voltages.

In order to boost the gain of telescopic OTA, the gain required for the feedback amplifiers is nearly equal to 50db. We can see that UGF of auxiliary amplifiers must be greater than 3-db of the main amplifier. It must be greater than 1.6 MHz. Schematic of the folded cascode OTA used for the Vbn or nbias is shown in Fig. 4.12. It is also called as the bottom auxiliary amplifier.

W/Ls of the folded cascade OTA used for Vbn and Vbp are calculated in the similar way by assuming the overdrives and from the current equation as shown for the telescopic OTA. M1 is used to bias the folded cascode OTA. M0 and M4 form a n-channel input differential pair. The two legs next to the input pair is a cascode load. The other half is a CMFB circuit. As already discussed its use and design in telescopic OTA, here it is used to maintain the outputs at Vbn which is 2.5 V in this design.

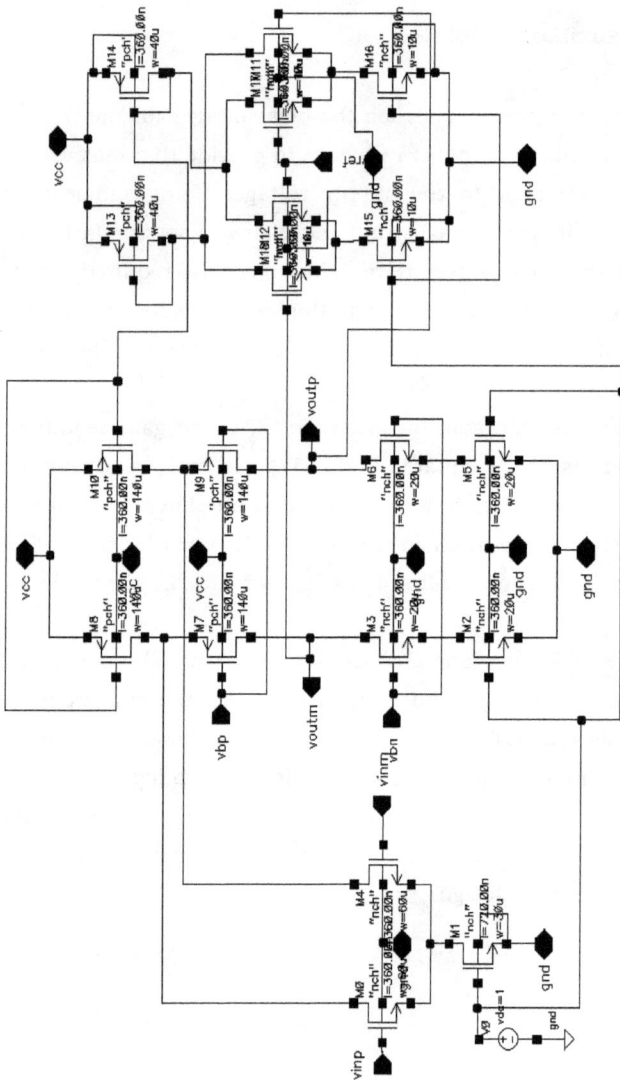

Fig.4.12 Schematic of Folded Cascode OTA used for nbias (Bottom)

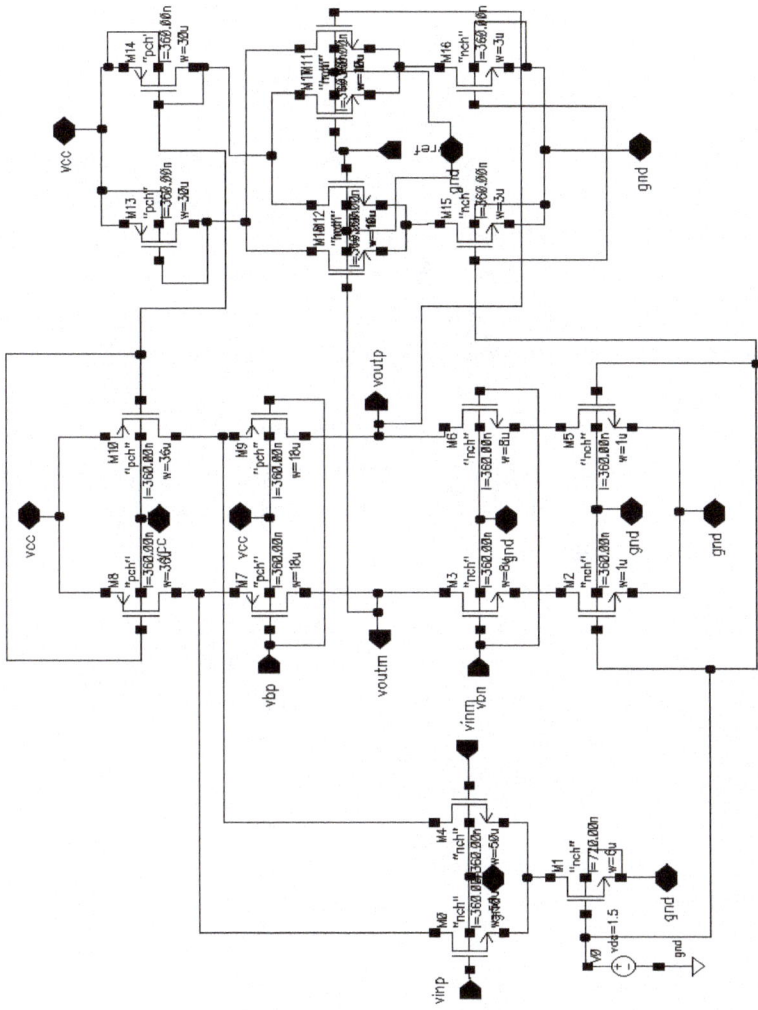

Fig.4.13 Schematic of Folded Cascode OTA used for pbias (Top)

Similarly the schematic of the folded cascode OTA used for the Vbp or pbias is shown in Fig.4.13. It is also called as the top auxiliary amplifier.

4.5.4 Design of Telescopic OTA with gain boosting

Gain boosted OTA designed by the three blocks discussed so far. These are the two folded cascode auxiliary amplifiers and the main telescopic amplifier. These have been integrated together to form Gain boosting architecture to provide high gain and high speed. The designed schematic is shown in Fig.4.14. The outputs of two folded cascode OTA are connected to the main amplifier as shown to provide the bias and the required gain boosting.

To ensure the stability of the system compensation capacitors have been used in the auxiliary amplifiers. These provide the phase margin larger than 60 degree. The values of these compensation capacitors have been decided after a few trails i.e. 0.4pF.

The aspect ratios W/L of the transistors are varied from the theoretically calculated values to ensure that the transistors are function in saturation region. The W/L ratios are shown in the circuit itself beside each transistor.

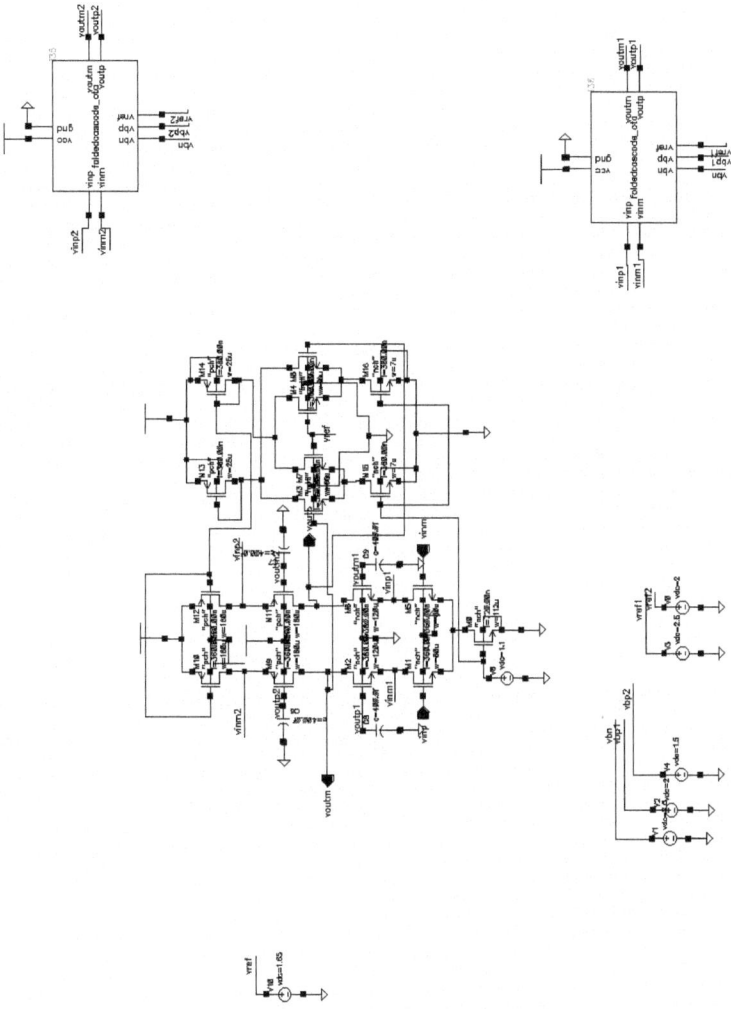

Fig.4.14 Schematic of Gain boosted telescopic OTA

91

4.6 Conclusion

A full analysis of the local common mode feedback OTA structure is presented. The addition of local common mode feedback was shown to have several performance improving benefits versus the conventional operational transconductance amplifier architecture including class AB operation which gives improvement in slewrate and gain bandwidth with equal static power dissipation. Implementation of local common mode feedback with MOS transistors MR1, MR2 is shown. It provides programmable gain (via the control voltage VR), allowing utilization of the same OTA for multiple applications. Class AB operation characteristics allow the local common mode feedback structure to outperform the conventional structure with unity mirror gain ($K=1$). MR1, MR2 provide the ability to enhance slew rate, gain bandwidth and phase margin. The benefits of the local common mode feedback structure are considerable for a negligible increase in area and equal static power dissipation.

Local common mode feedback OTA provides less gain and it cannot be used in A/D pipeline converters. So, Telescopic operational transconductance amplifier (OTA) is designed and to improve the gain auxiliary n-type and p-type Bias OTAs are designed. To achieve high gain and high UGF gain boosted amplifier is designed which is a combination of Telescopic OTA and Auxiliary OTA architectures.

CHAPTER 5

METHODOLOGY

5.1 Introduction

We have the six different CMOS operational amplifier architectures, or topologies, shown in Fig. 5.1-5.6.These are: A two stage operational amplifier as shown in Fig. 5.1, a simple Operational Transconductance Amplifier (OTA) as shown in Fig.5.2, an Operational Transconductance Amplifier (OTA) as shown in Fig. 5.3, a two stage cascode aperational transconductance amplifier as shown in Fig.5.4, a folded cascade operational transconductance amplifier as shown in Fig.5.5 and a telescopic operational transconductance amplifier as shown in Fig.5.6. Symmetry and matching constraints that must be met in many operational amplifiers. For example, the architecture input transistors should be match and the bias transistors should have the similar length to enhance the matching among them.

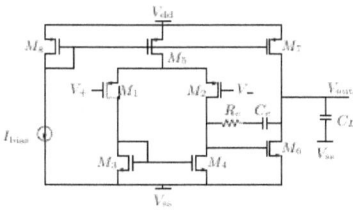

Fig.5.1 Two stage operational amplifier

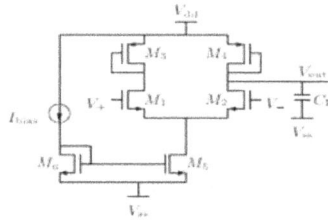

Fig.5.2 Simple OTA

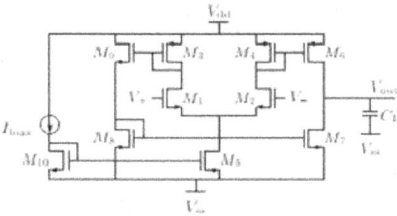

Fig.5.3 OTA

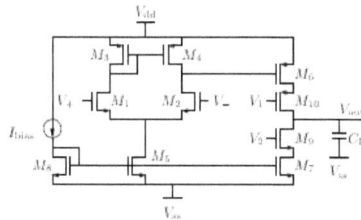

Fig.5.4 Two Stage Cascode OTA

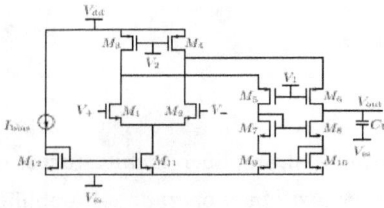

Fig.5.5 Folded Cascode OTA   Fig.5.6 Telescopic OTA

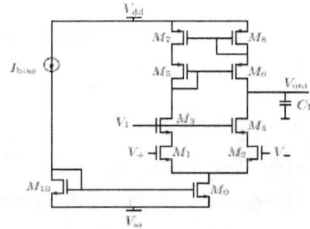

5.2 Design of CMOS operational amplifier steps

1. Select the design specifications and specify design inputs and outputs with proper functionality. This is the first step for designing any digital or analog circuits. Design fundamental architecture and their interconnections.

2. Calculate the specifications defined by designer with hand calculations, after representing the schematic.

3. In the software represent the schematic with proper interconnections and power supply. After check the circuit and simulate. Simulated circuit results to be stored with input and output values.

4. Compare the stored circuit input and output values with hand calculations. If both results are similar and meet the specifications not need to change the structure. Otherwise, modify the structure and follow the above step up to desired specification values.

5. Most of the efforts to be required selecting the dc currents in the schematic. Depend upon these transistor sizes (W/L) decided to get specification results.

6. Next step is the physical implementation of the schematic. These include Placement of components, Floor planning, Layout and Power supply pins.

7. In the schematic extract layout physical parasitic capacitance, inductance and resistance. Reduce these parameters and re-simulate the circuit with different input signals.

8. Schematic layout verify with original schematic representation.

In this thesis design of the Telescopic OTA and folded cascode OTA explained. Gain boosting circuit designed to increase the gain of the architecture. Two stage operational amplifiers provide high gain, high output swing, and medium power consumption. The major disadvantage is the speed. A few schematics were tried they are, either too sensitive to biasing or fail to meet the specifications. Folded cascode operational transconductance definitely can meet all the specification, like high gain, medium output swing, medium noise, high speed, good Power Supply Rejection Ratio (PSRR). Gain boosting configuration provides the superior gain at the cost of additional power dissipation.

Cadence software is used for the designed schematic and simulation of the specifications. Cadence is an Electronic Design Automation (EDA) environment tool that allows integrating in single framework different applications. Cadence supports all the stages of Integrated Circuits (ICs) design and verification. Cadence is completely general, supporting different fabrication technologies. In this design 0.18μm technology has been selected. These tools are used to design fully differential schematic view and also the layout circuit. Besides that, the simulation and results of the circuit can be obtained and can be analyzed via cadence. Circuit layout was created using the Virtuoso Editor in cadence. Design rule check then was carried out to make the layout design follow the rules of 0.18μm technology.

When a particular technology is selected, a set of configuration and technology related files are employed for preference the cadence environment.

For this design 0.18µm technology has been selected and all rules of this technology have been followed. Cadence provides the user to design analog IC design more easy as many application and tools can be applied. For example, Affirma Analog circuit design environment allows the user to do the circuit simulation. DC analysis, transient response and AC analysis can be done using certain steps in Cadence.

Library Manager: Analog and digital components are defined in the Library. Subsystems along with wires also included in the library. While designing any architecture designer has to pick the components and wires from the library.

Design Hierarchy: Placing instances of cell inside other cells forms the hierarchical design of the particular application or architecture. Top cell consist the entire design (ex: symbol of an inverter) and the bottom cell is the respected design layout. Level number zero defined for top cell and bottom cell as the highest level.

5.3 Library

Group of components, logical design objects, wires and sub components are included in library. In library basic design is the cell. Each cell includes one or more design cell views. Cell is defined logically according to physical design. In library virtual information is created according to cell and cell view. Give particular name to library. It forms an individual block for a design.

First Create a new library for the design in the library manager click on File then go to New after Library. Create new library particular name can be selected for the library. Left click the OK button. Next window will generate asking information detail about the technology file. Then attach to an existing tech file will be used. Click on the OK button.

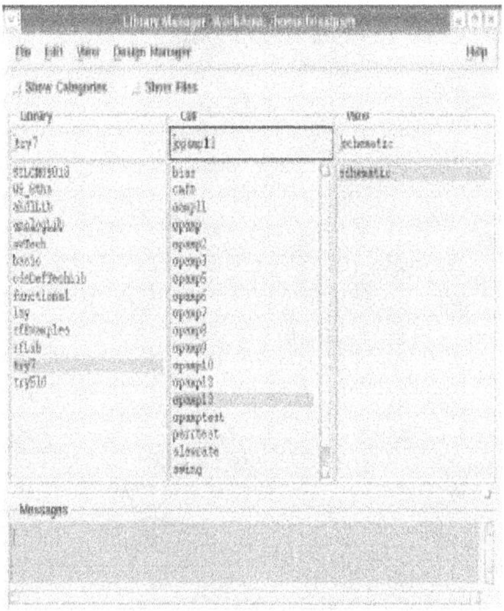

Fig.5.7 Library Manager

Next create a new cell click on library manager. In new go to cell view. This create new file form will appear on the screen. Give them any name in the cell view. In the view name block, type schematic or from the tool menu choose Composer-schematic and view name block will be automatically filled. Left click the OK button. The Virtuoso-Schematic editing window should be showing on the screen (big and black).

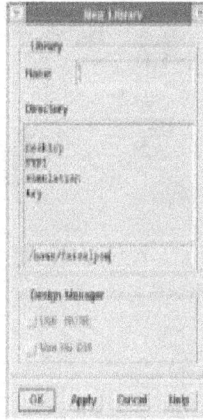

Fig.5.8 Creating new library dialogue box

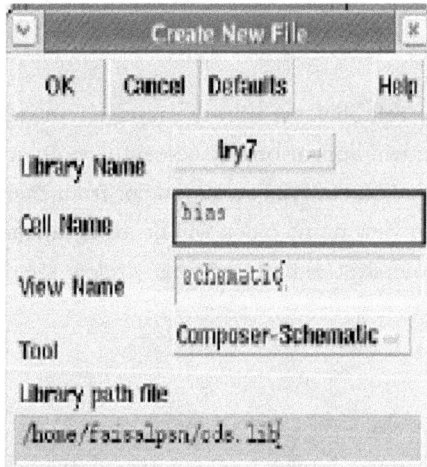

Fig.5.9 Creating new cell view dialogue box

Adding the Components: To add the components, left click on Schematic Editing: Add-> Instance or using 'I' button. For example, to add an n-type transistor, nhp in the SILCMOS018 library has to be used. Note that the browse button can be used in order to browse through the libraries and find the cell you want. Other than that, to add basic components such as ideal voltage or ground, analog LIb has to be used. To add the pins in the schematic, click on editing: Add-> Pin. The add pin dialog box will show up. In the pin names box enter the suitable pin name such as Vdd or Vss and ensure that direction for the pin is corrects either input or output. Then, to add wires, click on editing: Add-> wire (narrow). The add wire form will appear. Click its Hide button. Once the work editing had been done, left click on the "check mark" icon on the left side of the screen. This will check the work for connection errors and will automatically save the work in the library. This can be accomplishing using other method by left clicking editing: Design->Check and Save.

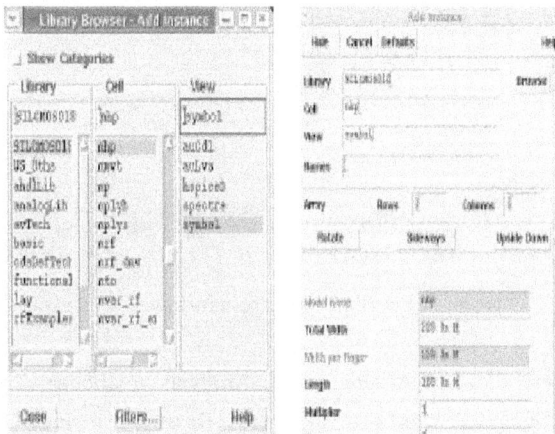

Fig.5.10 adding the components

Simulation with Analog Design Environment: The Affirma Spectre circuit simulator is a new architecture simulator. This tool used to simulate digital and analog architecture at the differential equation level. Spectre algorithms, the best presently available, give an enhanced simulator that is more accurate, faster, highly reliable, and more handy than previous Spice simulators. To do the simulation, firstly in the schematic, left click on editing: Tools->Analog Environment. The spectre window would appear.

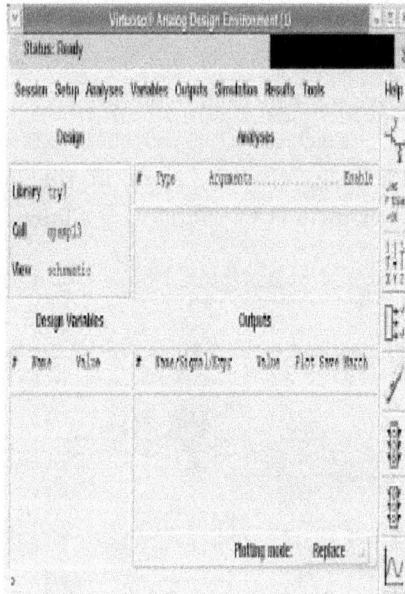

Fig.5.11 Spectre window

Firstly, left click on Analog artist: Analysis-> Choose. Then select the analysis that designer wants to perform. Example it is DC analysis, or transient analysis or AC analysis. Then, left click on Analog artist: Outputs->To be plotted-> Select on Schematic. Select the pin output in the schematic. Run the simulation by pressing on the screen traffic light icon and observe the results of the analysis. Now the result can be edited and printed.

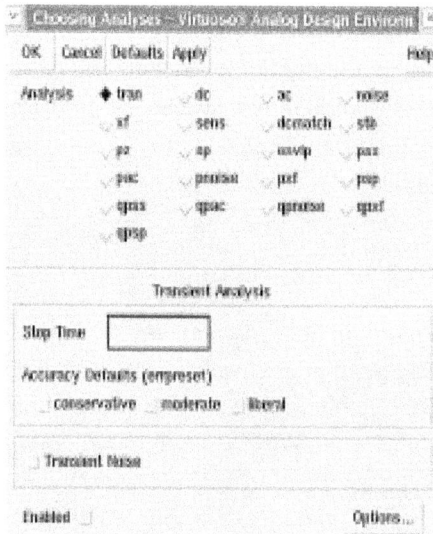

Fig.5.12 choosing analysis

In the designing we should know the spice models. Because creating new components one should have an idea about the spice models. In this thesis, the components needed in the analog designs only are explained.

5.4 PCB DESIGN

Designed circuit and bread boarded a functioning prototype now it is time to have a good Printed Circuit Board (PCB) design. PCB design for some designers, easy extension of the design process and it will be a natural. For many designers the way of circuit arrangement and deposit components/blocks on printed circuit board can be a very tedious work.

Appropriate printed circuit board design is very often an integral part of a design. In lot of designs like large speed and accuracy digital, RF analog and

102

low level analog to name a less the printed circuit board layouts may make or break the operation and electrical performance of the design. PCB foot prints consists resistance element, inductance element, and capacitance element, just like the architectures does. There are number of good practices to follow and basic rules, but apart from that printed circuit board design is a largly creative and individual process.

5.4.1 PCB Packages

Freeware, shareware or limited component full versions are some of the many PCB design packages that are available in the PCB design market.

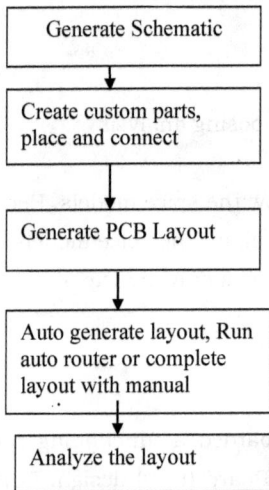

```
┌─────────────────────────┐
│   Generate Schematic    │
└───────────┬─────────────┘
            ↓
┌─────────────────────────┐
│  Create custom parts,   │
│  place and connect      │
└───────────┬─────────────┘
            ↓
┌─────────────────────────┐
│  Generate PCB Layout    │
└───────────┬─────────────┘
            ↓
┌─────────────────────────┐
│ Auto generate layout, Run│
│ auto router or complete │
│ layout with manual      │
└───────────┬─────────────┘
            ↓
┌─────────────────────────┐
│   Analyze the layout    │
└─────────────────────────┘
```

Fig.5.13 Flow chart of Layout design

The flow chart for layout design steps is shown and all these steps are explained as follows.

5.4.2 The Schematic

103

Generating a complete and accurate schematic diagram is the first important step for layout design. If the schematic is not accurate, layout design becomes tedious and it takes twice the design time. PCB design naturally is influenced by the original schematic as it is a manufactured version of the schematic. The generated schematic has to be neat, logical and clearly laid out to make PCB design simpler.

The various layers from the top of the board must be always transparent for PCB packages to work appropriately. For manufacturing or checking purposes layout bottom is used. PCB may consist many layers depend upon manufacture. Layers must be routed through proper power supply and ground. Through the board method allows designer to read text on the bottom layers as a mirror image. Not only the component used by the designer, also the PCB manufacturing process used to assemble the PCB size, it's shape and dimensions of pad. For various pad sizes and layout a whole slew of standards and process are developed.

Vias are used to connect the tracks by way as a hole in the board from one side as the board to another. Electrically plated holes called Plated through Holes (PTH). These are used to make vias in the board. Different layers on the board are electrically connected using PTH. Both via and pad practically are just electrically plated holes.

Pads and vias should be treated differently as these are differences when it comes to PCB design packages. Therefore, a pad cannot be used in place of via and vice-versa. Via holes is typically 0.5 to 0.7mm which is smaller than component pads. Stitching is the process of connecting two layers using a via. Since connecting two layers is analogous to electrically stitching both layers.

5.4.3 Component Placement and Design

PCB design comprises of ninety percentage of placement and ten percentage of routing. Placement is the very important step of the layout design. Efficient placement makes the PCB layout design job simple and best electrical performance can be obtained. Whereas un knowledgeable component placement leads to poor electrical performance. There is no right method defined for placement where every designer follows their own method of placement.

If different experienced designers are given the same circuit to do placement results different PCB layouts. It will be every time depends upon the placement of components. But a few basic rules are to be followed which provides easy routing, better electrical performances and simplify large and complex designs. Basic steps for complete board layout are:

1. According to the layout set the visible grid, snap grid, track size and the pad size.
2. Place all the required components onto the board.
3. Divide the components and place into the building blocks to make it proper functional.
4. First identify and route the critical layout tracks.
5. Off the board, each building block has to be placed and routed separately.
6. Completed building blocks are placed in main PCB board.
7. Signal and power connections are routed between blocks.
8. By optimizing the connections tidy-up of the board is done.
9. Proceed with design rule check.
10. Allow other designer to check it.

The required components are selected from the library by the designer and placed manually on the board, with which the designer gets a clear idea of whether the selected components easily fits onto the size of board being selected. If the design appears tight the designer reworks on the design and optimizes the component spacing and tracking. If enough room exists on the board the designer can be liberal in the layout design.

The Design Rule Check (DRC) is essential to ensure the correct connectivity between the components and functional blocks. Track connection between the components, correct width of the wires and clearness checked by DRC. Allowing other designers to recheck the board is very essential to get quality PCB.

5.4.4 Basic Routing and Power Planes

The other name of routing is the tracking. Routing is used to connect components by laying down tracks on the board. Net is the electrical connection between two or more pads which are to be kept as short as possible. Longer tracks will be associated with large parasitic resistance, capacitance and inductance.

The power wiring parasitic inductance and resistance to the components can be reduced drastically by the use of power planes. They are mostly used to distribute power along the board. This is considered to be good design practice. Even double sided boards can use power planes to distribute power, if many signal tracks are on the top layer. It is one solid copper layer dedicated to either ground or power rails or both. They can even go in the middle layers which are closer to outer surfaces of the board.

For signal reference a ground rail is used so it is preferred before a power rail. Special power plane layers are designed and laid in reverse to the other normal tracking layers in many PCB packages. Generally the designer lays tracks on the board which is blank to form copper tracks. On power plane board is assumed to be covered with copper where tracks are laid by removing copper.

5.4.5 Auto Routing and Auto Placement

Auto routing is a technique of PCB software to route the tracks of the designed circuit or entire board. This technique is not preferred by experienced PCB designers. For complex PCB boards which has less routing place on critical paths auto routers produce results much faster than human designer. Designers to route static control signals are tedious job example LED displays, switches and relays.

Auto placement tools are available in many high end PCB packages. Utilization of these tools is not preferred by expert designers. These tools provide an easy way to spread the components across the board. It is suggested for simple PCB board. Using auto placement software it is difficult to get optimization layout.

5.5 Conclusion

In this chapter different architectures are explained and elaborated design of CMOS operational amplifier steps. Basic information of Cadence tool is also explained. Component Placement and design PCB design flowchart, routing and placements are discussed.

CHAPTER 6

RESULTS AND APPLICATIONS

6.1 Introduction

Given architecture parameters are shown below in the table 6.1.a

S.No	Parameters	specifications
1	Technology	0.18µm CMOS
2	Power supply (Vdd,Vss)	±1.65V
3	Bias current (I bias)	500µA
4	Load capacitance (C_L)	55pF
5	V_{CMREF}	0V
6	V_R(SE-LCMFB)	-0.75V
7	V_R(FD-LCMFB)	-0.9V
8	Gain boosting architecture	2.5V –V_{bn} 2V –V_{bp}
9	Bias current (Gain boosting architecture)	1.2mA
10	Load Capacitance for gain boosting	0.4pF

Table 6.1.a Architecture Parameters

1. **Single Ended Conventional Architecture:** Single Ended Conventional Architecture is designed and simulated. The frequency response results the unity gain bandwidth of 11.5MHz with an open loop DC gain of 53db, 5.3V/µs slewrate with maximum output current of 0.48mA and settling time of 200ns. These parameters are not suitable for large frequency applications like A/D converters. This architecture dissipates static power of 4.91 mW, 15.9 mV of input offset voltage with an input noise at 10MHZ is 20µV.

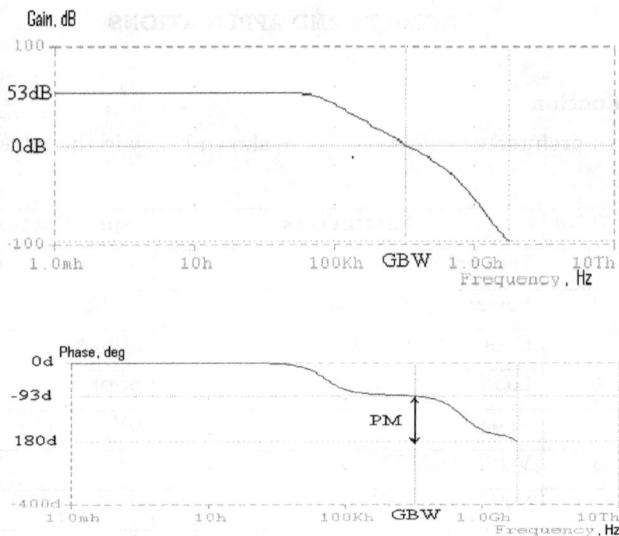

Fig.6.1 AC response of Single Ended Conventional Architecture

2. Single Ended Conventional with local common mode feedback Architecture: Single Ended Conventional with Local Common Mode Feedback Architecture is designed and simulated. The frequency response shows about 41.6MHz unity gain bandwidth with open loop DC gain of 67.8db, 20.7V/μs slewrate, 2.34mA of maximum output current with a settling time of 33.7ns. This architecture provides static power dissipation of 4.98 mW, Input offset voltage of this architecture (Vos) is 15.9 mv and input noise at 10MHz is 14μV. Clear improvement observed in the significant parameters compared to the conventional CMOS OTA at the cost of phase margin reduction. Therefore, it has been observed that significant parameters of local common mode feedback are better compared to conventional CMOS OTA.

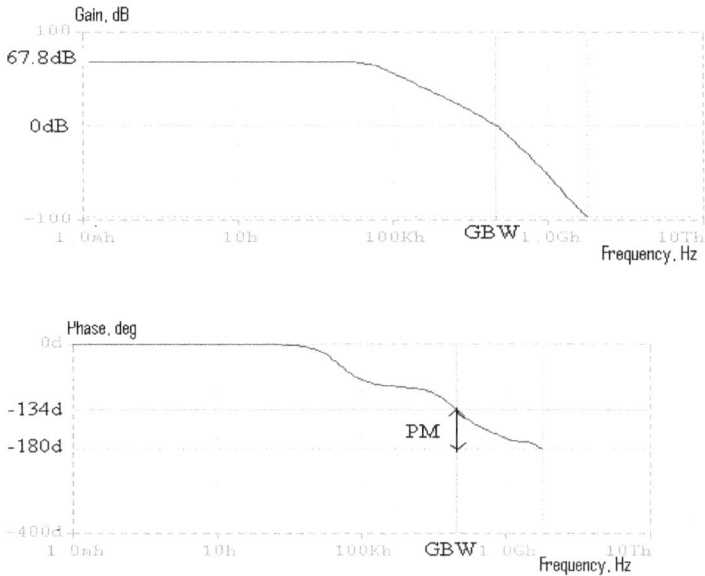

Fig.6.2 AC response of Single Ended Conventional with LCMFB.

3.Fully differential Architecture: Fully Differential Architecture is designed and simulated to provide an open loop gain of 62 dB, 160 MHz Gain bandwidth, 0.48mA of maximum output current and a lesser slewrate i.e. 15.1 V/µs. This architecture provides static power dissipation of 200 mW with 1.9 mV of Input offset voltage and an input noise at 10MHz of 18µV and Phase margin of 67 degree.

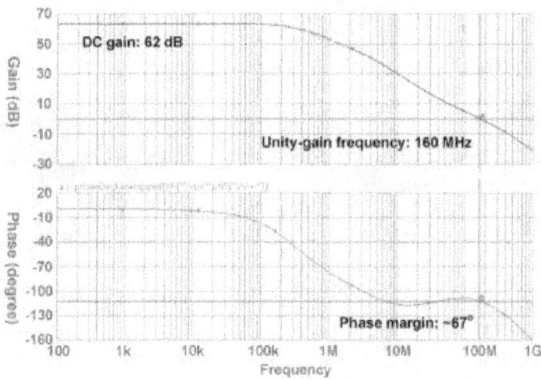

Fig.6.3 AC response of fully differential Architecture

4.Fully differential Architecture with local common mode feedback Architectures: Fully Differential Architecture with Local Common Mode Feedback Architecture is designed and simulated to provide the open loop gain (dB) of 67 dB, 270 MHz Gain bandwidth, 2.08mA of maximum output current. It provides less slewrate i.e. 9.92V/μs, dissipates static power of 4.91 mW, with gain margin (dB) and phase margin (degree) of 15.6 and 78.5 respectively. The Input offset voltage of this architecture (Vos) is 8.3 mV and input noise at 10MHz is 15μV.

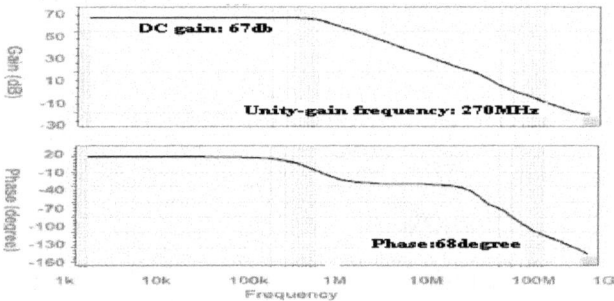

Fig.6.4 AC response of Fully differential Architecture with LCMFB

Above results indicate the effect of the application of common mode feedback, the reduction in gain and phase margin of both single ended and fully differential architecture LCMFB can be used to trade off slewrate and gain bandwidth enhancement with gain and phase margin (stability)

5.Telescopic OTA: Telescopic operational transconductance operational circuit is designed and simulated providing an Open loop DC gain of 60.7db, 1.8GHz of Unity Gain Frequency (UGF), 62^0 of Phase Margin (P.M) and 1.5V (p-p) of Output voltage swing.

AC Response

- phaseDegUnwrapped(VF("/net018")/VF("/net039"))
- dB20(VF("/net018")/VF("/net039"))

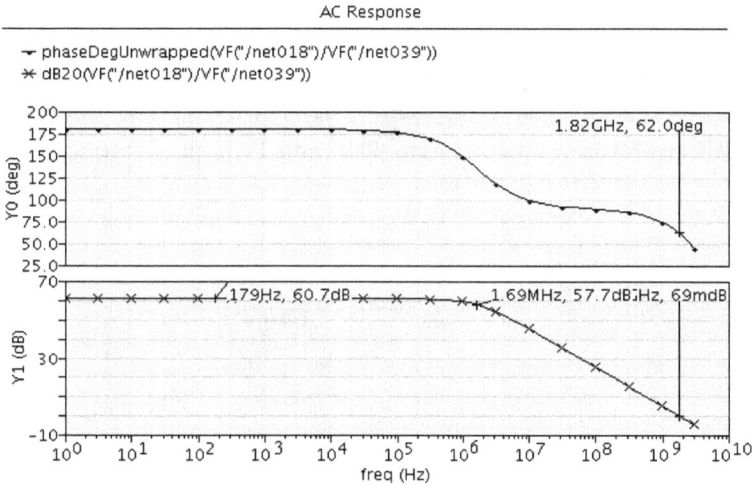

Fig.6.5 AC response of Telescopic OTA

6. Auxiliary n-bias OTA: This circuit is used for bias in Gain Boosting Architecture. This architecture provides 56.1db of Open loop DC gain, 733MHz of Unity Gain Frequency (UGF), 60.1^0 of Phase Margin (PM) and 1V (p-p) of Output voltage swing.

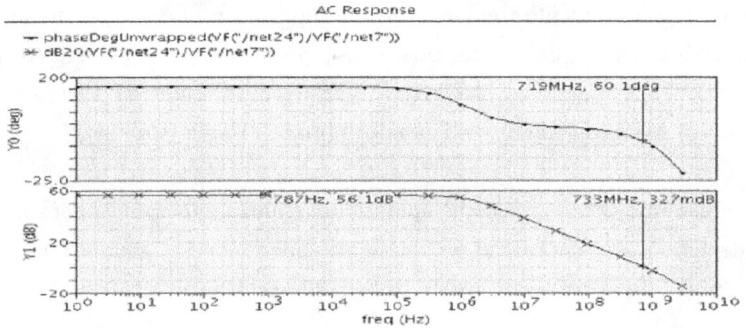

Fig.6.6 AC response of top auxiliary amplifier used for n-bias

7. Auxiliary P-bias OTA: This circuit is used for bias in Gain boosting circuit. This circuit provides Open loop DC gain of 62.2db, 692MHz of Unity Gain Frequency (UGF) , 59.3⁰ of Phase Margin (PM) and 1V (p-p) of Output voltage swing.

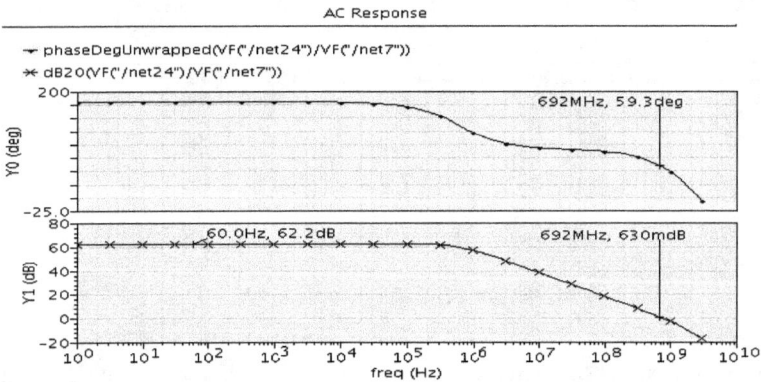

Fig.6.7 AC response of top auxiliary amplifier used for p-bias

8. Gain boosted OTA: Combination of Telescopic OTA, n-Bias and p-Bias becomes Gain Boosted OTA. This architecture provides Open loop DC gain of 110db, 1.8GHz Unity Gain Frequency (UGF), 62.4⁰ of Phase Margin (PM) and

1.576V (p-p) of Output voltage swing. This architecture is used for A/D application.

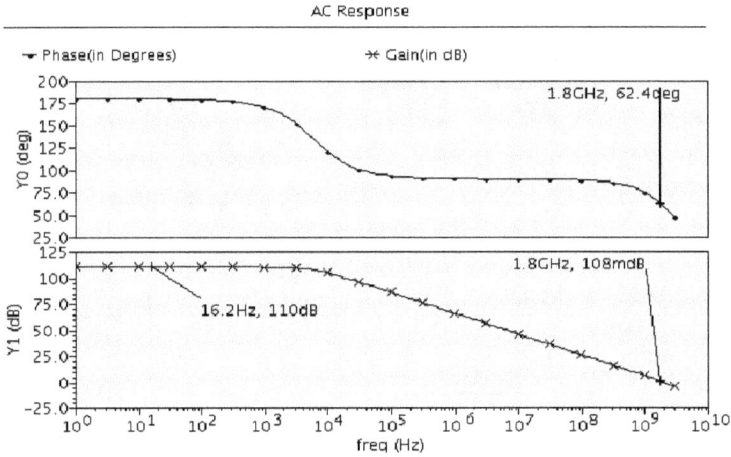

Fig.6.8 AC response of Gain boosted OTA

DC response of the Gain boosted Amplifier is shown in Fig 6.9. Output swing of the amplifier is 1.57V as appears in graph which is greater than 1V (from the specifications)

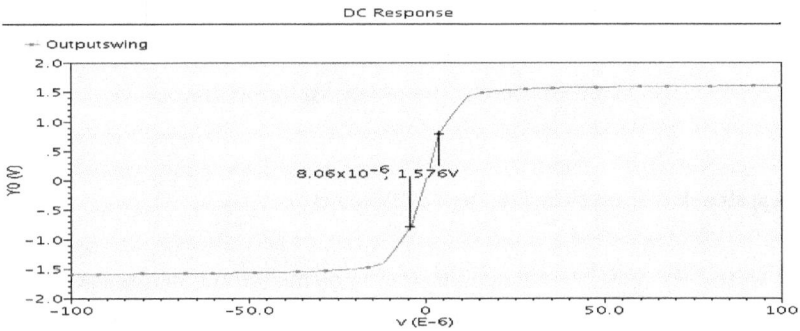

Fig.6.9 DC response of gain boosted telescopic OTA

Designed Gain Boosted OTA architecture compared with various researches paper. The comparison is shown in Table 6.1.b

S.No.	Parameters	Gain Boosted OTA	Beth Isaksen [33]	K.Bult [34]	K.Gulati [35]	Hwang [36]	Nordiana [37]
1.	No. of Stages	1	1	1	1	2	1
2.	Gain (dB)	110	80	90	90	77	93.27
3.	GBW	1.8 (GHz)	1000 (MHz)	116 (MHz)	90 (MHz)	870 (MHz)	9.32 (MHz)
4.	Load Capacitance (pF)	0.4	1	16	4	1	4
5.	PM (degree)	62.4	72	62	78	56	93.14
6.	VDD (V)	2.5	3	5	3.5	3	5
7.	Power Cons. (mW)	18	17	52	15	15	16
8.	Slewrate (V/µs)	20	18	18	22	15	36
9.	Process (µm)	0.18	0.25	1.6	0.8	0.25	0.5

Table 6.1.b Comparison of Architectures

6.2 Application for Analog to Digital Converter

Research has been done on the implementation of analog to digital converters. A number of techniques for doing analog to digital conversion have been developed. In this chapter, some of the standing out architectural styles are introduced and compared. Each architecture has its own advantages and disadvantages, and each has a set of applications for which it is the best solution.

6.3 Flash ADC

The Flash ADC architecture is fundamentally the fastest architecture. It's also known as a fully parallel architecture. Flash ADC n bit consists of an array of 2^n-1 comparators and a set of 2^n-1 reference values. Every comparators samples the applied input signal and compares the signal to one of the given reference values. Each comparator output indicates whether the input signal is greater or lesser than the given reference assigned to that comparator. Comparator outputs are often referred to as a thermometer code.

A simple 3 bit flash ADC is shown in fig.6.10. The encoder converts the thermometer code produced by the comparators to a binary code as shown in the truth table in table 6.2. As seen from the figure, the comparators all operate parallelly. Thus, the conversion speed is controlled only by the speed of the comparator or the sampler. For this reason, the flash A/D converter is capable of large speed.

Flash A/D converter requires the large hardware and 2^n-1 comparators. It is sensitivity to comparator offsets and dissipates more power. The large numbers of comparators present a large capacitance to the output of the sampling circuit. The required comparator offset voltage for a flash A/D converter with n bit resolution is less than $\frac{1}{2^n}$. Required comparator offset becomes very small at high resolutions. Comparators with less offsets are difficult to design and costly. To get small offset many comparators are required. ADCs with higher resolutions than eight (8) bits frequently use the flash circuit.

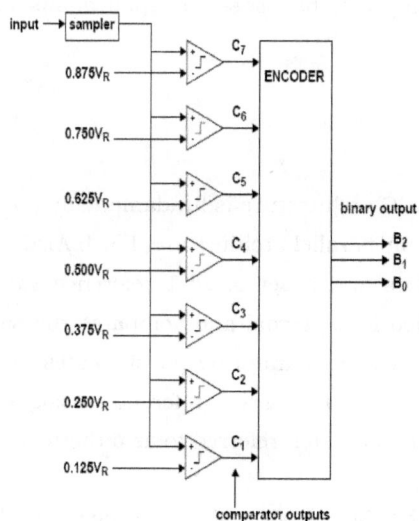

Fig.6.10 Simple 3-Bit Flash ADC

Table 6.2 Input/output Table of Flash ADC

	C_7	C_6	C_5	C_4	C_3	C_2	C_1	B_2	B_1	B_0
$V_I < 0.125V_R$	0	0	0	0	0	0	0	0	0	0
$0.125V_R < V_I < 0.25V_R$	0	0	0	0	0	0	1	0	0	1
$0.25V_R < V_I < 0.375V_R$	0	0	0	0	0	1	1	0	1	0
$0.375V_R < V_I < 0.50V_R$	0	0	0	0	1	1	1	0	1	1
$0.50V_R < V_I < 0.625V_R$	0	0	0	1	1	1	1	1	0	0
$0.625V_R < V_I < 0.75V_R$	0	0	1	1	1	1	1	1	0	1
$0.75V_R < V_I < 0.875V_R$	0	1	1	1	1	1	1	1	1	0
$V_I > 0.875V_R$	1	1	1	1	1	1	1	1	1	1

6.4 Two Step Flash ADC

117

The structure of a two step flash analog to digital converter is shown fig.6.11. The conversion does not happen all at once as in the flash A/D converter. Here, the conversion requires two steps. During the first step, the most significant bits of the digital output are determined by the first stage N/2 bit flash analog to digital converter called Coarse A/D converter. Then a D/A converter convert this digital result back to an analog signal to be subtracted from the input signal. This residue is then sent to the second stage N/2 bit flash A/D converter called Fine A/D converter. The second stage flash then resolves the least significant bits of the digital output.

The conversion time for a two step flash A/D converter is longer than for a simple flash, but it is still very fast. Furthermore, the two step flash ADC requires only $2 * 2^{n/2}$ comparators, which are fewer comparators than required by a simple flash A/D converter. For instance, if the 10 bits are total resolution, the first period can be quantized the first 5 MSB's and the next period is the next 5 MSB's. In each period only 5 bits are quantized, the need number of comparators is about 2^5 in each period, and the total comparators are $2x2^5 = 64$ as aganist to 1024 in the straightforward 10bit flash analog to digital converter. Like this considerable amount of power can be reduced at the cost of an extra clock cycle. Thus, the two step flash A/D converter saves hardware. As a result, two-step flashes A/D converters are often used in the 10 bit resolution range.

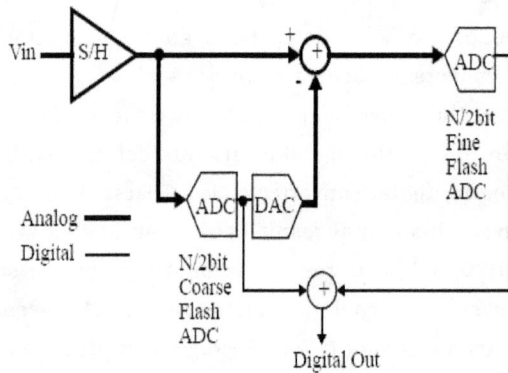

Fig.6.11 Block diagram of 2 step flash architecture

6.5 Two Step Flash ADC with Interstage Amplifier

By including an interstage gain amplifier fine comparators accuracy requirements can be relaxed to amplify the signal for the fine comparator bank as shown in Fig 6.12. Here, the gain of $2^{\frac{n}{2}-1}$ is compulsory used instead of $2^{\frac{n}{2}}$ in order to remove the over range problem, and the fine flash resolution of A/D converter is increased by 1.

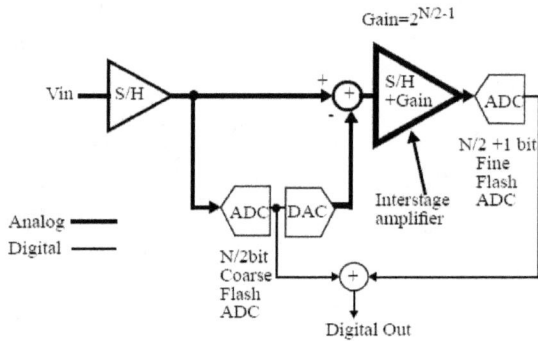

Fig.6.12 Block Diagram of 2 step flash ADC with an interstage amplifier

One more benefit of this architecture is that the conversion steps can be pipelined due to the sample and hold inter stage amplifier while the first stage flash analog to digital converter works on the most recent sample, the second stage flash can at the same time work on the previous sample.

Sampling and quantization's are need only 2 clock periods. In this architecture the throughput can be enhanced. However, an operational amplifier must be considered for the sample and hold gain block and its power can be significant if fast output settling is needed. With the use of a sampling capacitor input sample and hold function can be included in the comparator, the inter stage amplifier must be carryout with a SC circuit which usually needs an operational amplifier. It has to drive $2^{\frac{n}{2}+1}$ comparators in the fine flash A/D converter section; the operational amplifier dissipates more power if N is high.

6.6 Subranging ADC

A subranging ADC architecture is a multistep converter architecture that includes two-step flash ADCs and also includes ADCs that extend the concept of the two-step flash ADC to a larger number of steps. By breaking the conversion process into multiple steps it requires less comparator. But the conversion time is longer. Fig.6.13 shows the structure of the subranging analog to digital converter. Each stage is compelled for resolving some part of the digital output word and conveying a residue to the following stage. The conversion time required increases with the number of stages while the hardware required decreases with the number of stages. Thus, there is a arrangement between hardware and speed. The comparators in the front stages need not be accurate, but the comparators in the last stage must be accurate to the full resolution of the analog to digital converter.

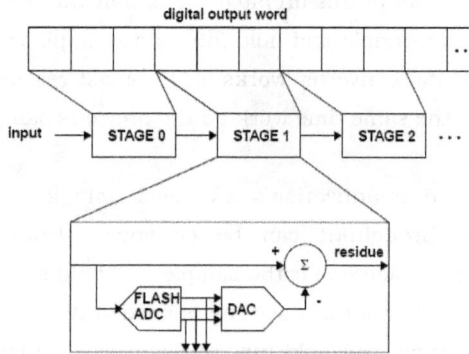

Fig.6.13 Sub ranging ADC

6.7 Pipelined ADC

6.7.1 Introduction to the Concept of Pipelined ADCs

A pipelined A/D converter is a better example of a pipelined signal processor. One example of A/D converter is a pipelined sub ranging A/D converter. The conversion process is split into a number of steps in a sub ranging A/D converter. A certain number of bits of the digital outputs is analyzed during each step of the exchange. In the first step the most significant bits are analyzed, and in the last step the least significant bits are analyzed. A pipelined sub ranging A/D converter is a sub ranging A/D converter which has a processor dedicated to each step of the conversion operation. In other words, a pipelined A/D converter contains a number of stages. Each stage of the pipeline is answerable for resolving some segment of the digital output word. In a pipelined A/D converter, the various stages of the pipeline operate concurrently. For example, as 3rd stage measure sample 1, stage 2 measure samples 2, and stage 1 measure sample 3. Pipelined A/D converter is another type of sub ranging ADC that has features that improve the throughput rate and tolerance to comparator errors.

121

In Fig. 6.14 a basic schematic for pipeline architecture is shown. A previous stage signal samples by each stage and it quantizes to B+1 bits by the flash analog to digital converter. Then, the quantized signal is subtracted and the residue is amplified through the interstage amplifier to be sampled by the subsequent stage. The same operation is done in each stage down the pipeline to perform analog to digital conversion. The number of comparators required in this case is the number of stages times. From Fig. 6.13, it is roughly $(M*2^{B+1})$. The required number of stage is approximately the analog to digital converter resolution divided by effective per stage resolution. B is denoted as the effective per stage resolution, and digital correction required one extra bit. Due to the interstage gain and digital correction blocks flash ADC required B+1 bit. Comparator requirement is depends upon the B.

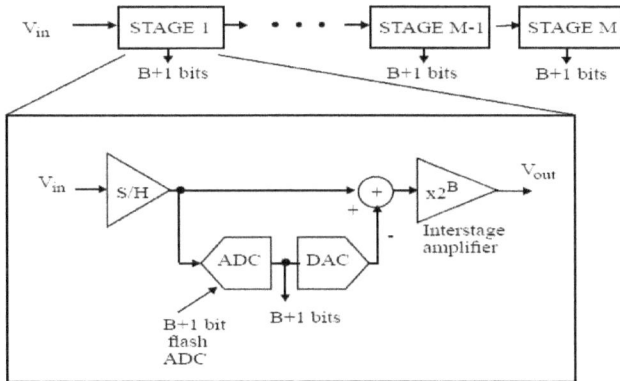

Fig.6.14 A Typical Pipeline Architecture

A typical pipelined A/D converter is a subranging converter having the basic comparison, subtraction and amplification block repeatedly. The block diagram of a conventional pipelined A/D converter is shown in Fig.6.15. As shown, a pipelined A/D converter has one block every time the processing step is repeated. Every block also includes sample and hold circuit to hold the analog applied signal or residue signal. This feature allows the first stage of the pipeline to perform a coarse quantization on a sample of the signal while the second stage processes the last sample. In this converter without pipelining, all the steps in quantizing a signal must be completely finished before the next sample can be taken. In a pipelined A/D converter, a higher throughput rate can be obtained because a new sample can be considered as soon as the first stage of the pipeline has finished processing the old sample.

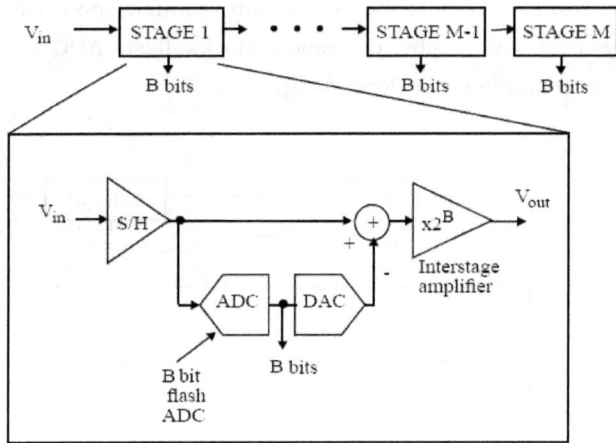

Fig.6.15 Block Diagram of a conventional Pipelined ADC

Although the throughput rate is independent of the number of stages in the pipeline, conversion time for any given sample is proportional to the number of stages in the pipeline. This is true because the signal must work its way through all the stages before the complete output word is generated. This delay can be an issue if the pipelined A/D converter is part of a feedback system. An amplifier is used to amplify the residue before passing it on to the next stage. One significant assumption of this is that the comparators in the last stages of the pipeline need not be accurate to the full analog to digital converter resolution as they are required to be in other subranging analog to digital converters.

The disadvantage of adding the gain blocks is that they aim to be the dominant source of power dissipation in the A/D converter. Therefore, pipelined analog to digital converters tend to dissipate high power than subranging analog to digital converters. However, like the other subranging analog to digital converters, pipelined analog to digital converters can achieve high resolutions with relatively low hardware.

Both interstage amplifier and digital to analog converter requirements get relaxed down the pipeline. If the A/D converter resolution is 10bit and B=1, while the first stage has to meet 10bit requirement, the need on the second stage is relaxed by 1bit. The numbers of comparators are further reduced. But it increased latency and required sample and hold circuits.

6.7.2 Digital Error Correction

Digital correction of 2-bit pipeline stage is explained with an example. Fig 6.16 shows the block diagram of a 2-bit pipeline stage. As discussed here for a 2-bit pipeline stage the interstage amplifier has a gain of $2^2=4$. The ideal transfer function of a 2-bit stage is shown in Fig.6.16. The possible outputs of ADC are 00, 01, 10 or 11. The range of input and the output here for a 2-bit ADC are from $-V_{ref}$ to $+V_{ref}$. The transfer function shown in Fig.6.16 is drawn assuming the Sub-ADC and interstage amplifier as error free.

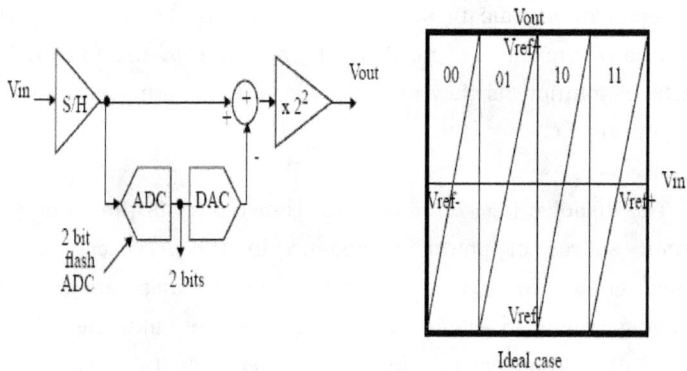

Fig.6.16 Block Diagram of a 2-bit Pipelined Stage with its Ideal Transfer Function

The digital output for the stage is increased by one bit, when the input crosses one of the sub-ADC decision levels, whereas the output stage decreases by $2V_{ref}$. The signal presented to the next stage with the interstage gain of 4 is the full scale and with out errors in sub-ADC and Sub-DAC. When an offset error occurs in the sub-ADC comparator or in the interstage amplifier the output of the first stage will exceed the range as shown in Fig.6.17. Then saturate the second stage and leads loss of data. To overcome this problem, increase the range of the second stage sub-ADC or equivalently reduce the interstage gain of the first stage to tolerate sub-ADC error [7].

O in above figure Indicates Over range Error. When the interstage gain is reduced to 2, the transfer function becomes as shown in Fig.6.17. This allows the sub-ADC error to be as high as $V_{ref}/4$ and the output are still in the input range of the following stage. However, when a sub-analog to digital error is present without digital correction, the error will appear in the final digital output. In other words, if digital correction is not used, the first stage sub-ADC must still be as linear as the entire converter. Because of interstage gain the requirements can be relaxed. Now, assume the first stage is ideal, with a full-scale input to the first stage, the output is only between $-V_{ref}/2$ and $+V_{ref}/2$,

125

leading an extra bit on top and bottom of the per-stage resolution. Digital correction simply utilizes the extra bit to correct the over ranging section from the previous stage.

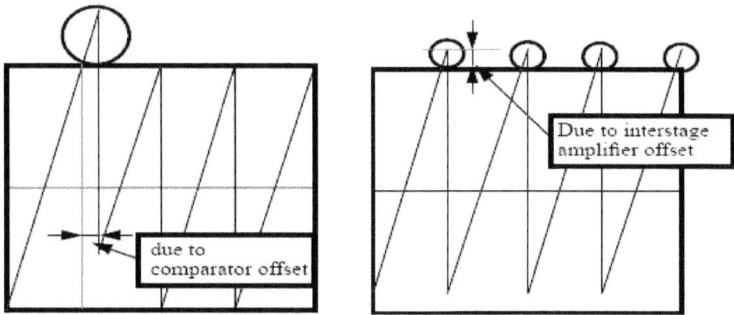

Fig.6.17 Transfer Function with Comparator Transfer Function with Inter stage Offset error amplifier error

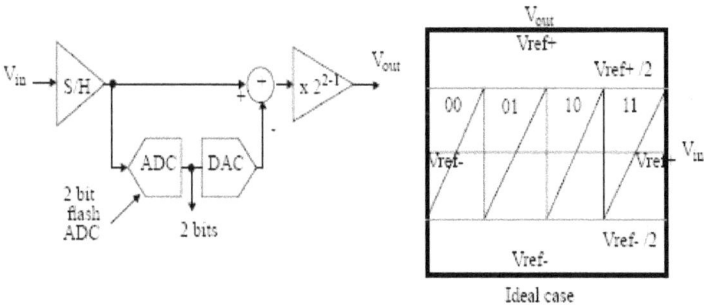

Fig.6.18 Block Diagram with reduced gain and Ideal Transfer Function.

For example, when one of the sub-ADC thresholds has an offset, the output of the first stage exceeds by $+V_{ref}/2$ as shown in Fig 6.18. The second stage, sensing the overranging, increases the output by 1 LSB.

126

This bit causes the first stage output to increase by 1 LSB during the digital correction cycle. In the same way, when the output of the first stage drops below $-V_{ref}/2$, the second stage will sense the overranging and subtracts 1 LSB during digital correction cycle. With this method, the sub-ADC error, as large as $V_{ref}/4$, in the stage can be corrected by the following stage with digital correction.

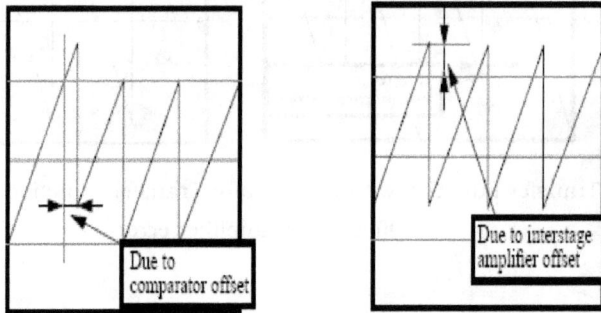

Fig.6.19 Transfer functions with comparator   Fig.6.20 Transfer function with interstage.

Offset error amplifies error with above digital correction algorithm; both addition and subtraction need to be present in the digital correction circuit which complicates the code assignment for the pipeline stage. Subtraction can be eliminated by intentionally adding an $-V_{ref}/4$ offset to the sub-ADC and the output of sub-ADC. A conceptual block diagram and transfer function is shown in Fig. 6.21. With this configuration, the sub-ADC error, upto $V_{ref}/4$, can be tolerated and digital correction circuit is modified to contain adders only.

127

Before adding 1/2 LSB offset

After adding 1/2LSB

without top comparator

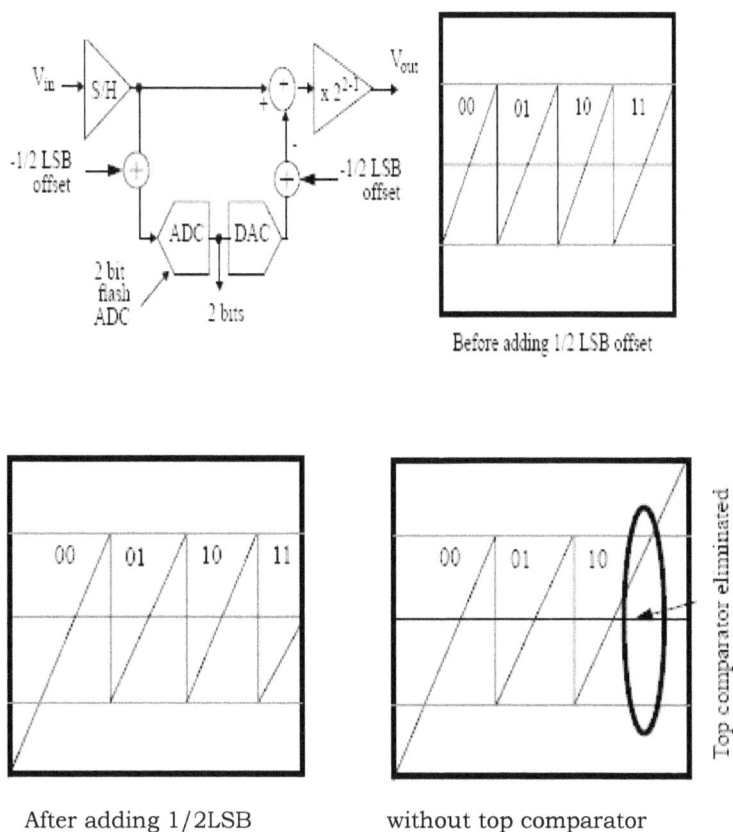

Fig.6.21 Conceptual block diagram with digital correction

Since over ranging in the transfer function can be detected by the next stage, one can simplify design even more by eliminating a comparator at $3V_{ref}/4$ as shown in Fig. 6.21. The final block diagram and transfer function is shown in Fig. 6.22. As shown now it can be called as 1.5-bit stage instead of 2-bit ADC because although the output of a single stage is 2-bit, the LSB bit in the former stage is used to correct the error using digital error correction. For the 1.5-bit stage comparator thresholds are at $V_{ref}/4$ and $-V_{ref}/4$. The sub- DAC levels are at $-V_{ref}/2$, 0 and $V_{ref}/2$. The codes are shown on top of the transfer function and the over ranging part on the transfer function will be digitally corrected by the next stage except the last stage of the pipeline.

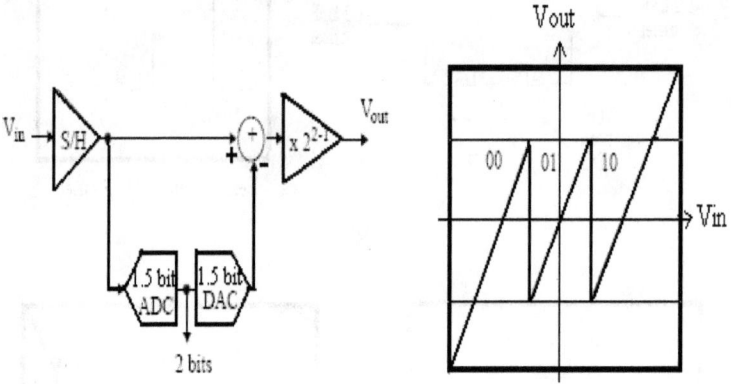

Fig 6.22 Block diagram of a 1.5-bit stage

But in this design a 2.5-bit pipelined stage has been chosen. The concept behind the 2.5-bit is similar to the one explained in the above sections with 3-bits instead 2-bits. For the 2.5-bit stage the gain of the interstage amplifier is $2^{3-1}=4$. The comparator thresholds are at $-5V_{ref}/8$, $-3V_{ref}/8$, $-V_{ref}/8$, $V_{ref}/8$, $3V_{ref}/8$ and $5V_{ref}/8$. The Sub-DAC levels are at $-3V_{ref}/4$, $-2V_{ref}/4$, $-V_{ref}/4$, 0, $V_{ref}/4$, $2V_{ref}/4$ and $3V_{ref}/4$ for the output codes 000, 001, 010, 011, 100, 101 and 110 respectively.

	Ideal	Reduced gain With
	(normalized w.r.t V_{ref})	modified coding

129

Input Range	-1,1	-1,1
ADC Threshold Level	-3/4,-1/2,-1/4,0, 1/4,1/2,3/4	-5/8,-3/8,- 1/8,1/8,3/8,5/8
DAC Levels	-7/8,-5/8,-3/8,- 1/8,1/8,3/8,5/8,7/8	-3/4,-1/2,- 1/4,0,1/4,1/2,3/4
Interstage Gain	8	4

Table 6.3 Comparison of 3-bit/stage and 2.5-bit/stage

From the discussion so far a table 6.3 that compares the original coding and digital error correction with modified coding for a 3-bit stage. To get the output code of the whole pipeline ADC, original coding just combines the codes from each stage one next to the other. Five stages are needed for 10-bit output code. If digital error correction is applied, the output code is obtained by overlapping one bit between neighboring stages (as shown below). Five stages need 10-bit output code.

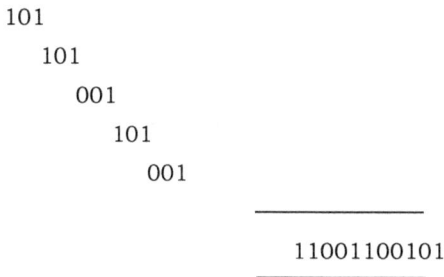

```
101
 101
  001
   101
    001
   _____

    11001100101
   _____
```

As shown above a simple shift and addition is performed to get the required digital output. So here we get a latency of 5 clock cycles (depends on no. of stages).

6.8 Comparators

The basic function of a comparator is to provide sufficient gain so that the digital output levels can be attained in response to small difference between two input signal levels. Three generic approaches to design the comparator are the single pole amplifier (SPA), the multistage amplifier (MSA) and the regenerative sense amplifier (RSA) as shown in Fig.6.23.

The regenerative sense amplifier (RSA) was chosen in this project for sub-ADC due to its better performance than others. The major performance of a comparator is determined by speed, power and resolution (offset).

Let the initial input voltage = 0 and a step input is applied at t=0. Then the output voltage Vo will grow exponentially as t increases as shown in Fig.6.23 (a), (b) and (c), where Ao is the steady state gain. After an amplification time, U is defined as an equivalent amplification and as a ratio of the output Vo to the input step amplitude Vi and can be considered as a comparison parameter.

For the RSA, instead of the above definition U, an equivalent amplification U of the RSA is redefined as the ratio of the differential output Vo1 (t)-Vo2 (t) to the initial imbalance Vo1 (0)-Vo2 (0) because of the positive feedback action to the initial imbalance. In RSA architecture because of the positive feedback action minimum t_a can be achieved.

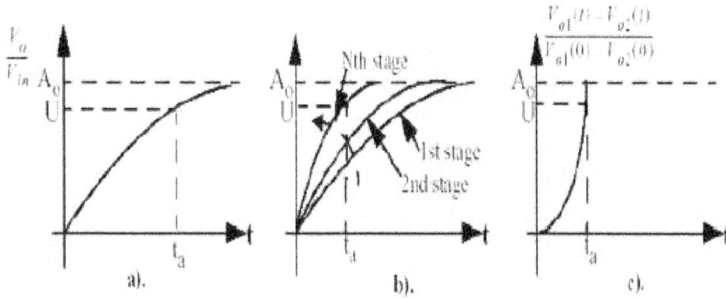

Fig 6.23 (a), (b), and (c) give the gain as a function of time for SPA, MSA and RSA respectively

For the power comparison, we assume the power dissipation in each stage to be approximately same. The lowest power dissipation amplifier is SPA, the largest dissipation one is MSA because of the requirement of several numbers of stages. The RSA dissipates moderate amount of power. But the overall performance of the comparator is evaluated by the power-delay product. In first order analysis, the power dissipation of each stage is Pm so P(SPA) = Pm, P(MSA)=NxPm and P(RSA)=2Pm.

It can be seen that the minimum power delay product is obtained for RSA. Therefore, from the above analysis, it can be concluded that the RSA architecture is better for high speed and low power comparator.

RSA has comparatively large offset. Since the digital error correction algorithm is used in our pipelined ADC, the offset requirement of sub-ADC is relaxed, the comparator threshold voltage in sub-ADC can be varied within some swing in a 2-bit per stage pipeline ADC. Hence the use of low power dynamic latch as a comparator is feasible even if it inherently has a comparatively large offset than the general differential amplifier.

132

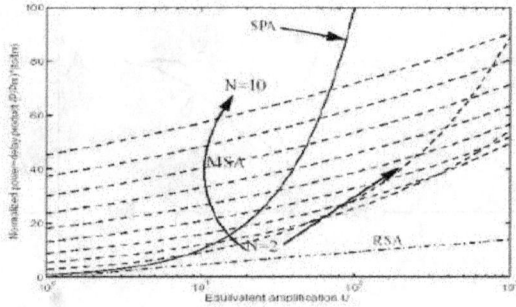

Fig 6.24 Normalized power-delay product of three comparator architectures

6.9 Dynamic latched type comparator

The schematic of the latch-type comparator is shown in Fig.6.25. The operation of the comparator begins when the V_{latch} is low; the two output nodes (Von & Vop) pre charge fully to digital 1 level (3.3V) and the lower NMOS cross-couple circuit is disconnected from the upper part by the NMOS switches in this phase. When the V_{latch} is high, the upper pre charging switches open and stop pre charging, the lower NMOS discharging circuit starts its operation.

The comparison is made by comparing the rate of discharging the output node (Von & Vop). When Vin+ input has a higher voltage than another input, the rate of the voltage drop at node Vop is faster than that at node Von due to the larger current flow at M1. This generates an imbalance voltage which is further enhanced by the regenerative action.

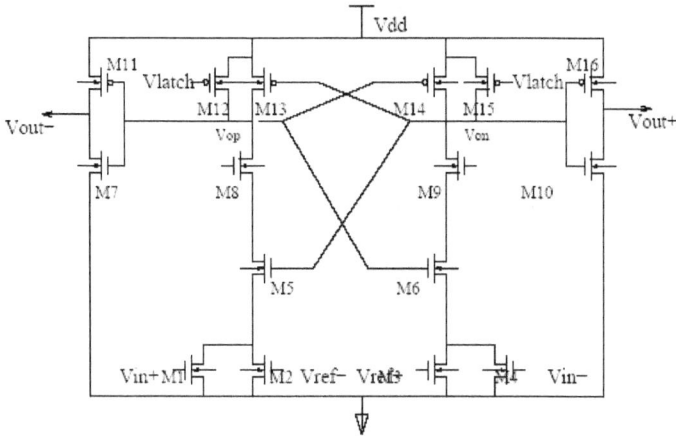

Fig 6.25 Dynamic latched type Comparator

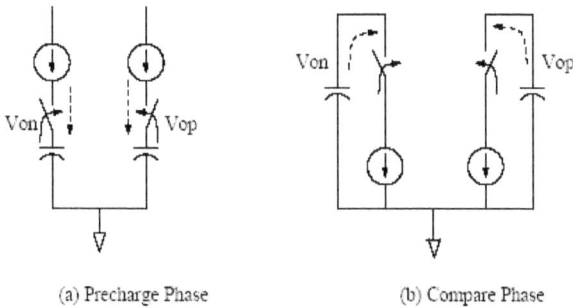

(a) Precharge Phase   (b) Compare Phase

Fig 6.26 The simplified Principles of the operation of the comparator

6.10 Design of different blocks used in PIPELINED ADC

This chapter also describes the design of various blocks used in pipelined ADC such as Clock Generator, Sub-ADC, MDAC or Gain stage and Sub-DAC.

All the blocks are simulated individually AT 100 M-Hz and the results are shown. The block diagram of the pipelined ADC used in this design is shown in Fig 6.27.

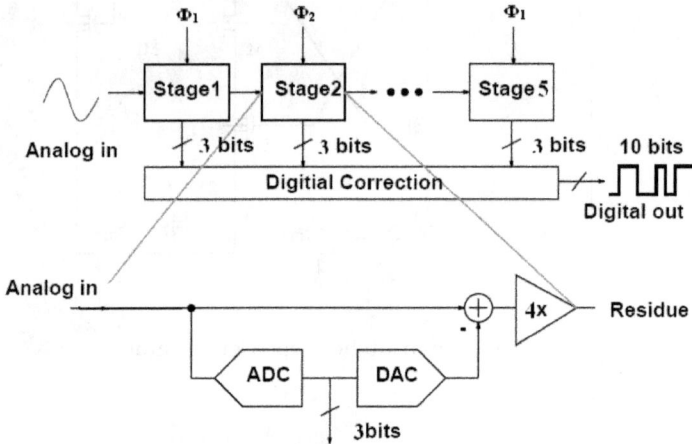

Fig6.27 Block Diagram of Pipelined ADC used in this design

As can be seen it consists of 5 stages of which odd stages operate on Φ_1 and even stages operate on Φ_2. The 3-bit outputs from all the stages should go to the shift register (not shown in figure) and then to the Digital Error Correction Circuit to get the final 10-bit Digital Output.

6.10.1 Clock Generator

All stages of pipeline ADC rely on a two phase non-overlapping clock for operation. All the odd stages sample during phase phy1 (Φ_1) and present a valid residue output to the next stage during phase phy2 (Φ_2). Even stages work on the opposite phases. All the stages operate at the same time. The two phases, Φ_1 and Φ_2 have a 180 degree phase shift and a delay between the clock transitions. In Switched Capacitor circuits the bottom plate switch is opened first to reduce signal-dependent charge injection. So the additional clock signals Φ_{1d} (phy1d) and Φ_{2d} (phy2d) are also generated. Φ_{1d} and Φ_{2d} are designed to turn off before Φ_1 and Φ_2 respectively. The circuit designed to generate the clocks is shown in Fig.6.28. It consists of NAND gates, Inverters and DELAY element.

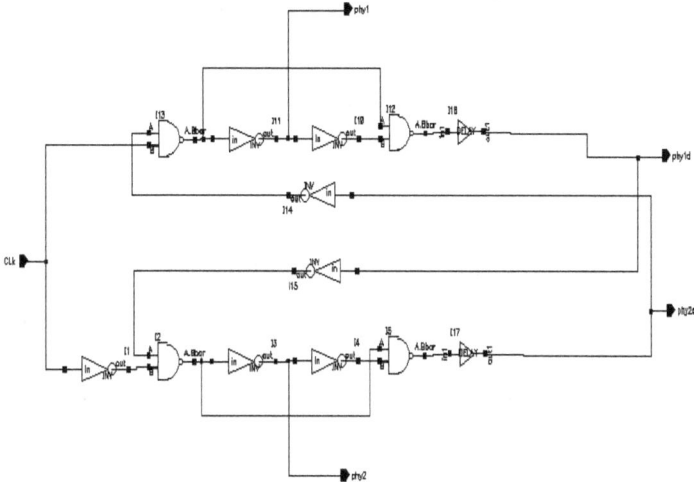

Fig.6.28 Schematic of Clock Generator

Delay element shown in above figure consists of four minimum sized inverters cascaded together and was used to adjust the duration time of four clock phases. All clock outputs were buffered to increase the drive capability.

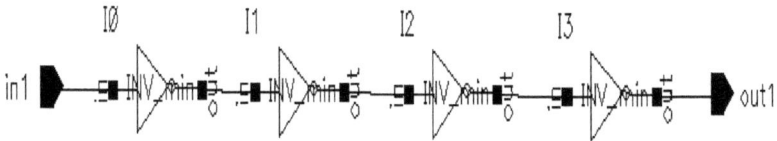

Fig.6.29 Schematic of DELAY element

The simulation results of the clock generator at 100M-Hz is shown in Fig.6.29. As can be seen after Φ_{1d} takes transition from HIGH to LOW, then only Φ_2 takes transition from LOW to HIGH. In this period the Sub-ADC outputs have to settle and provide a constant output to the DAC during Φ_2.

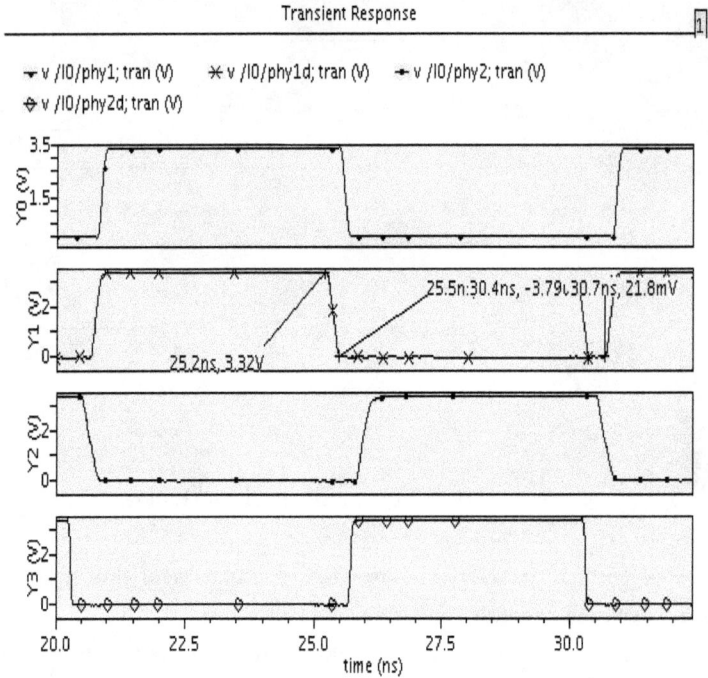

Fig.6.30 Clock Generator simulated at 100M-Hz

6.10.2 Sub-ADC

The function of Sub ADC is to quantize the input signal and provide the intermediate bits for each stage. The 2.5 bit per stage architecture can have one of the six binary states as an output: 000, 001, 010, 011, 010, 101 and 110. The 3-bit digital output of the Sub-ADC goes to the correction logic circuit to correct any errors occurred due to the comparator offset using the digital outputs of next stages.

Differential Input	Sub-ADC Output	B2 B1 B0

The block diagram of Sub ADC consists of 6 comparators (7th comparator is not needed because of correction logic) numbered from 1 to 6 with their thresholds set at 5Vr/8, 3Vr/8, Vr/8, -Vr/8, -3Vr/8, -5Vr/8 respectively. At the output of the comparators we got a 6-bit code in the format of series of 1's followed by series of 0's from MSB to LSB, called Thermome. So at the output as shown in the block diagram a Thermometer code to Binary code converter is required.

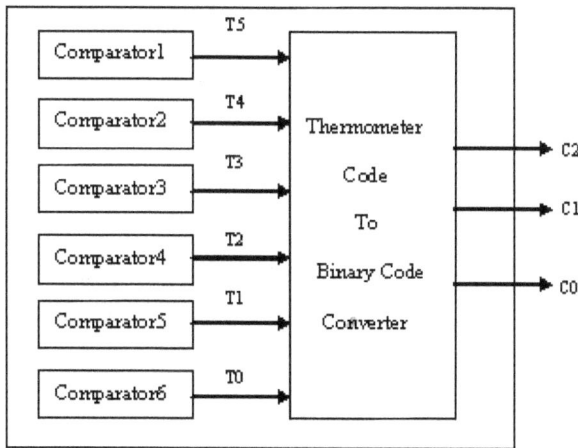

Fig.6.31 Block Diagram of Sub-ADC

138

Vin<-5Vr/8	000000	0	0	0
-5Vr/8<Vin<-3Vr/8	000001	0	0	1
-3Vr/8<Vin<-Vr/8	000011	0	1	0
-Vr/8<Vin<Vr/8	000111	0	1	1
Vr/8<Vin<3Vr/8	001111	1	0	0
3Vr/8<Vin<5Vr/8	011111	1	0	1
Vin>5Vr/8	111111	1	1	0

6.4 Input/output table of Sub-ADC.

Table shows the outputs of the comparator and the 3-bit digital output for the all possible combinations of differential input range (from $-V_{ref}$ to $+V_{ref}$). In this design it is from -0.5V to +0.5V. So the comparator thresholds from 1st comparator to 6th comparator are 312.5mV, 187.5mV, 62.5mV, -62.5mV, -187.5mV and -312.5mV.

6.10.3 Comparator

Dynamic latched comparators have been used to reduce the power dissipation. Although the offset error is high, it can be taken care by the Digital Error Correction circuit. The circuit diagram of the dynamic latched type comparator is shown in Fig 6.32.

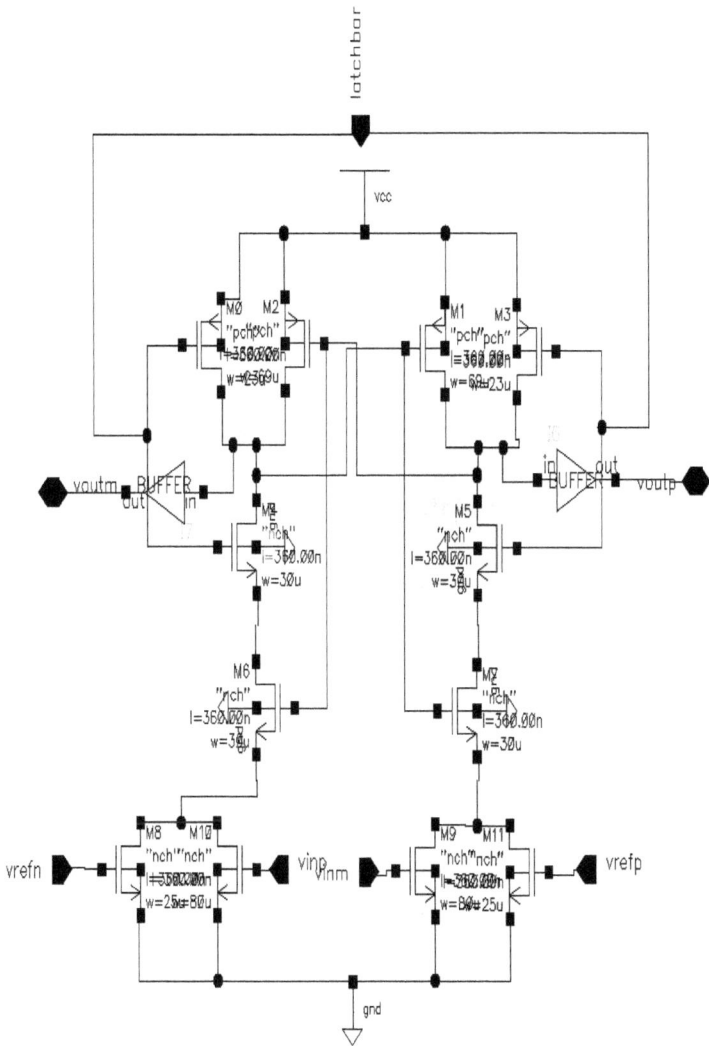

Fig.6.32 Schematic of Dynamic latched comparator.

Design parameters for the latch-type comparator

The most important part of this latch-type amplifier is the cross-coupled pairs during comparing phase and the ratio of the width of the driving transistors (M9 & M10) to that of the reference transistor (M8 & M11) (see Figure 4.6). The sizes of the cross-coupled pairs mainly determine the maximum speed of the comparator because of regenerative action. But the limitation is that the width (W) of the cross-coupled pair cannot be infinitely large to increase speed, because Cgd1 and Cgd2 are proportional to W and an increase in W increases the Miller capacitor Ceq due to the rapid rising of the differential voltage (Vgs2-Vgs1).

The minimum channel length (L) and width (W) should be used in the cross-coupled pair to maximize the speed performance, but the sensitivity of this pair is inversely proportional to the width (W).The ratio of current in two chains can be adjusted through the ratio of the width of M9 & M10 transistors to that of M8 and M11 transistors for the reference voltages. The W/L ratio of the reference transistors (M8 & M11) can be determined by considering the Transconductance of the lower part of NMOS transistors in linear region.

$$G_1 = \frac{1}{R1} = K_P \left[\frac{W_{in}}{L} (V_{in+} - V_{th}) + \frac{W_r}{L} (V_{ref+} - V_{th}) \right] \qquad (6.3)$$

$$G_2 = \frac{1}{R1} = K_P \left[\frac{W_{in}}{L} (V_{in-} - V_{th}) + \frac{W_r}{L} (V_{ref+} - V_{th}) \right] \qquad (6.4)$$

$$V_{in,threshold} = \frac{W_r}{W_{in}} . V_{ref} \qquad G1=G2 \qquad (6.5)$$

$$V_{in} = V_{in+} - V_{in-} \qquad (6.6)$$

$$V_{ref} = V_{ref+} - V_{ref-} \qquad (6.7)$$

Where W_{in} and W_r are the width of the input transistors (M9 & M10) and the reference transistors (M8 & M11) respectively. $V_{in/threshold}$ is the threshold of the comparator. The minimum length 0.18μm was used for the whole comparator for a high speed application and the widths were optimized to achieve the required performance. As already discussed, in sub-ADC we have 6 comparators with their thresholds set at 5Vr/8, 3Vr/8, Vr/8, -Vr/8, -3Vr/8 and -5Vr/8. So for the comparator1

$$\frac{5Vref}{8} = \frac{W_r}{W_{in}} \times Vref \tag{6.8}$$

$$\frac{W_r}{W_{in}} = \frac{5}{8} \tag{6.9}$$

Similarly for comparators 2 and 3 the ratio Wr/Win is 3/8 and 1/8 respectively. To get the negative thresholds for the remaining 3 comparators, V_{refp} and V_{refn} should be interched. In this design width of the input transistor Win is as small as 10μm. Since it is the first block in each 2.5-bit stage, if small value is chosen then the load capacitance for the previous stage driving is small. If large widths are chosen the output of the previous stage does not settle to the required value within the given period. W/Ls of the input and reference transistors for all the comparators are shown in Table 6.5.

Comparator	Input Transistor (μ/ μ)	Reference Transistor (μ/ μ)
Comparator 1	10/0.18	6.25/0.18
Comparator 2	10/0.18	3.75/0.18
Comparator 3	10/0.18	1.25/0.18

Table 6.5 W/Ls of the Comparators used in Sub-ADC

When latchbar in the circuit is 0 the outputs are precharged to 3.3V. Now if latchbar is 1 the output of the comparator is according to the input at which latchbar signal takes transition from 0 to 1. If that input is greater than threshold of the comparator it remains at 3.3V only i.e., output is logic1; if input is less, then output signal takes a transition from 3.3 V to 0 i.e., logic 1 to logic 0.

Here the comparators output will goes to the thermometer to binary code converter. So the comparators must be able to drive the load capacitance (sum of the drain capacitances of comparator and the input capacitance of code converter). But actually here they are not able to drive in this design, so buffers have been used at the output of the comparators.

The thermometer code values and their respective binary codes are shown. From this using k-map logic for B2, B1 and B0 is derived and the circuit shown above is designed. In this circuit AND gates are used at the starting to make output Logic '0' when latch is 0 and output is valid during latch is high.

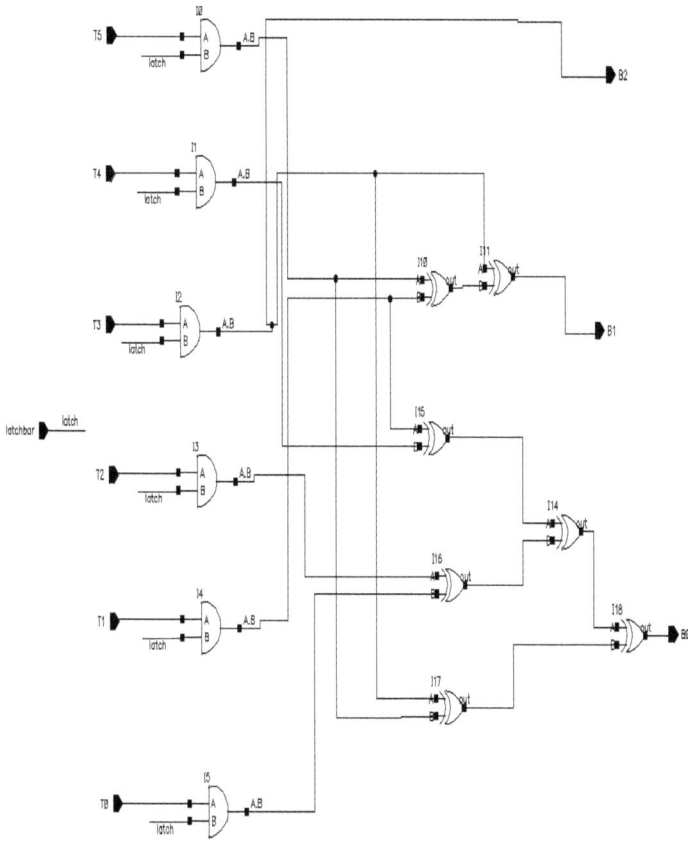

Fig.6.33 Thermometer to Binary Code Converter

The complete schematic of sub-ADC is shown in fig.6.34 with Clocks. The latch signal shown in thermometer to binary converter is not the same signal that has applied to the comparator. From fig. 6.34 it clearly shows that it is delayed clock with respect to latch. The comparators output will take some time to settle to the 0 values from 3.3V, so the signal OE (output Enable) is a delayed one. Sub-ADC has been tested with a ramp signal as the input and fig. 6.34 shows the output of the ADC corresponding to the given input.

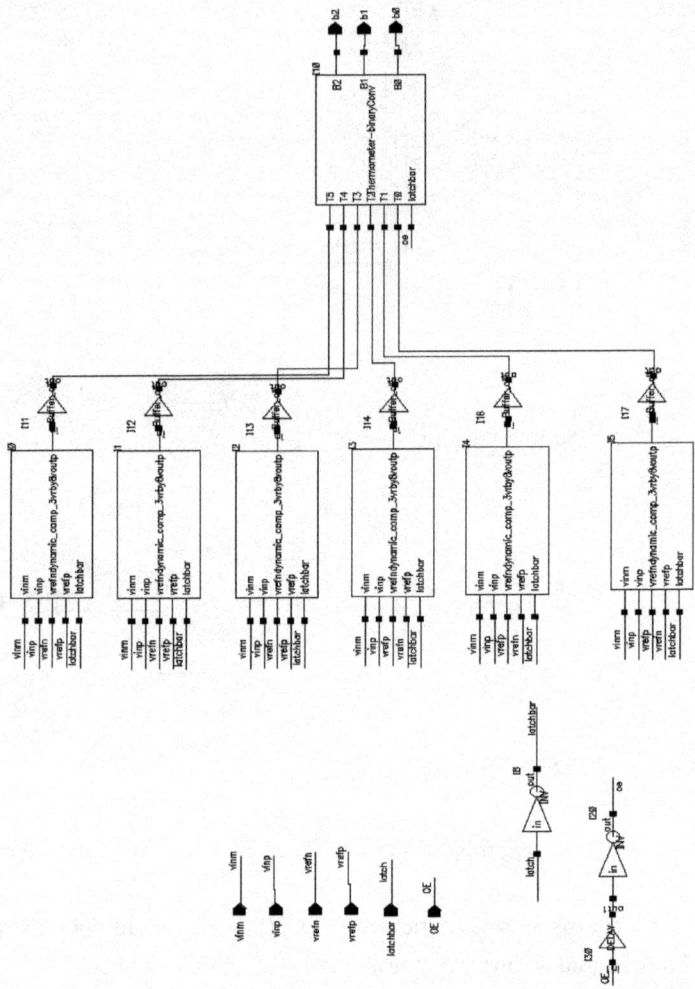

Fig.6.34 Schematic of Sub-ADC

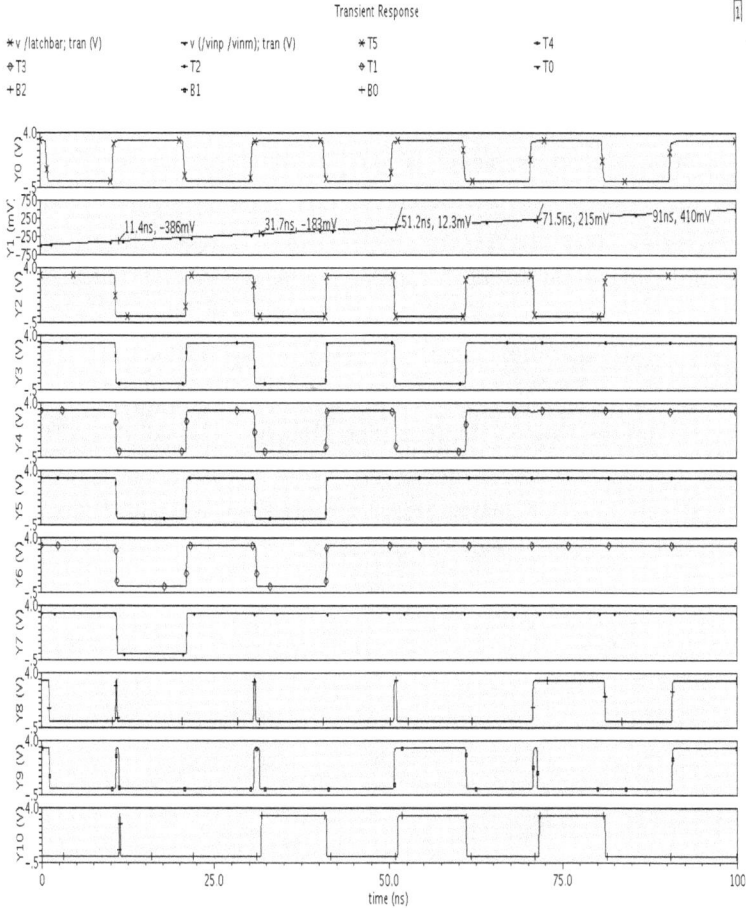

Fig.6.35 Response of sub-ADC to a ramp input for 100M-Hz clock (latch)

6.11 MDAC

This example illustrates one way of implementing the MDAC and residue amplifier, a major component of the pipelined ADC and also the critical block or the main block. As we know for a 2.5 bit stage, the residue amplifier has a gain of four. The function of this circuit is, first to S/H the input signal, next to produce a residue that is the difference between the input and some reference, third amplify this residue. In this approach, the circuit operates on two phases, a sampling phase and a hold phase. During the sampling phase shown in figure 6.36(a), the input signal is sampled onto the capacitors C1 and C2. During the hold phase the capacitors are then switched to one of seven voltages, $+V_{ref}$, $+2V_{ref}/3$, $+V_{ref}/3$, 0, $-V_{ref}/3$, $-2V_{ref}/3$, $-V_{ref}$. The voltage is chosen based on the digital output of the analog to digital converter block (Sub ADC).

As the voltage is switched, the input voltage to the large gain amplifier, also known as the summing node voltage, contributes to change. As it does, the output of the large gain amplifier changes a great deal. The negative feedback through the capacitor C_F turns this summing node voltage to zero. The result is that the charge initially stored on capacitors C1 and C2 is transferred to the capacitor C_F.

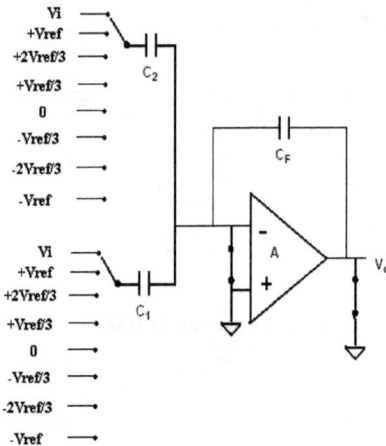

Fig.6.36 Sampling Phase

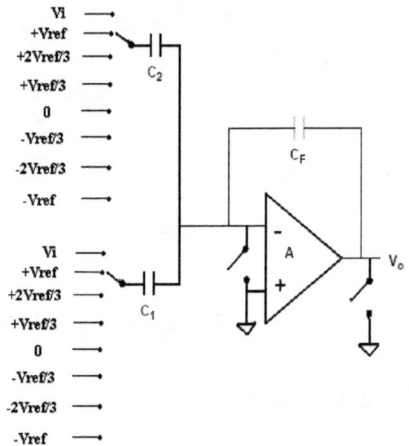

Fig.6.37 Hold Phase

Fig.6.36 and Fig.6.37 Operation of Switched Capacitor Sample/Hold Block with DAC, subtraction, and Residue Amplifier Included

For the case shown in the figure; the output voltage is a function of the input voltage and reference voltage.

$$V_O = \left(\frac{C_1+C_2}{C_F}\right)V_1 - \frac{C_1}{C_F}V_{ref1} - \frac{C_2}{C_F}V_{ref2} \qquad (6.10)$$

$V_{ref}1$ could be $+V_{ref}$, $+2V_{ref}/3$, $+V_{ref}/3$, 0, $-V_{ref}/3$, $-2V_{ref}/3$, $-V_{ref}$ depending on where C1 is connected during the hold phase. $V_{ref}2$ is determined in a same manner. A modified version of this switched capacitor circuit is shown in figure 6.38. In this scheme, the feedback capacitor is used for sampling during the sampling phase since it would normally be idle at this time anyway. The modified circuit has the benefit that it uses less capacitors. Because there are less capacitor at the input of the amplifier, the feedback, and thus the speed, is improved.

Disadvantage is during the hold phase, the switch in series with the feedback capacitor may degrade the settling speed. This decline may cancel the improvement from the improved feedback factor. The output voltage of this modified circuit is given by the above equation.

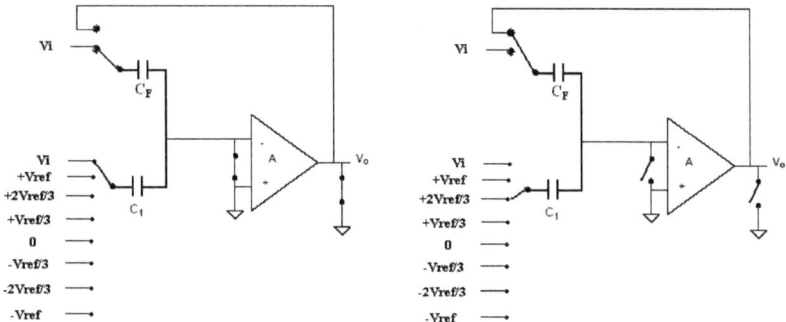

(a)   Sampling Phase   (b) Hold Phase

Fig 6.38 Operation of Switched Capacitor Sample/Hold Block with
Shared Feedback Capacitor

Here we may have a doubt that in the previous section we have seen that the sub DAC levels are at -3Vr/4, -Vr/2, -Vr/4, 0, Vr/4, Vr/2 and 3Vr/4. But here it is mentioned that V_{ref1} could be $+V_{ref}$, $+2V_{ref}/3$, $+V_{ref}/3$, 0, $-V_{ref}/3$, $-2V_{ref}/3$, $-V_{ref}$. This is because in this design, gain of 4 stage is not used at the output of every stage. For the 2.5-bit stage designed in this project, C_1 and C_F values chosen are 3C and C respectively.

$$V_0 = 4V_I - 3V_{Ref1}$$

So the actual outputs for the sub DAC from the theory and the outputs that obtained in this design are different. But ultimately the same residue that should be passed to the next stage from the theory is obtained. As mentioned above MDAC or the Gain stage is the critical block of pipelined ADC. In MDAC, OTA is the main block. Already design of OTA has been discussed in chapter 3 and 4. From chapter 4 gain boosted telescopic OTA is used as the OTA in this design.

For an odd stage, Φ_1 is used to sample and quantize the input signal and during Φ_2, the residual is generated by the gain stage and passed to the next (Even) stage. In Fig during Φ_1 the capacitors C1 are charged up to values of differential input signals. During this time, the differential outputs of the OTA are tied to common mode VCM. Once Φ_2 is active, the residual is created by subtracting DAC outputs from the input signal and this difference is sent to the next stage. Sizes of switches are optimized so that they can perform well at 100 MHz

149

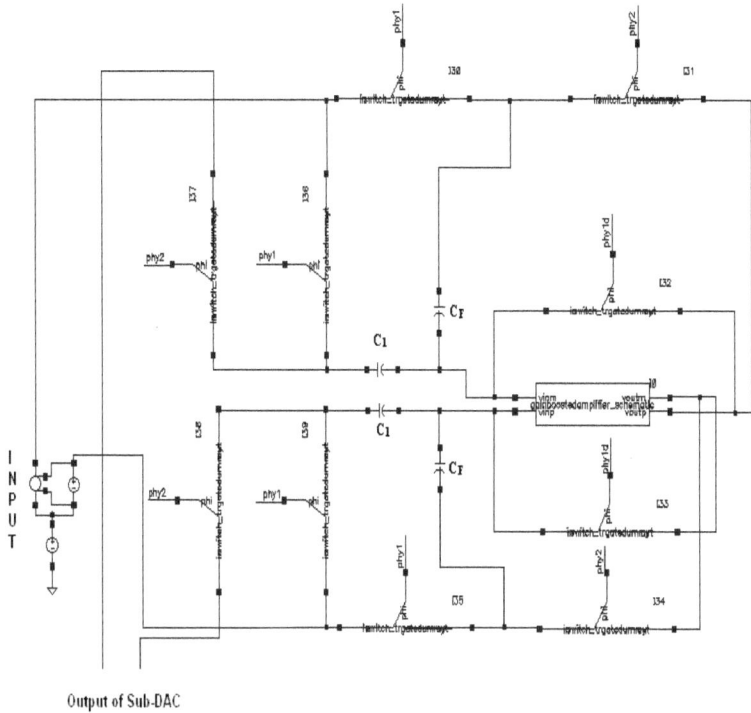

Fig.6.39 Schematic of MDAC

As shown in fig. 6.39 it operates on two clock cycles Φ_1 and Φ_2 which are out of phase. During Φ_1 both the capacitors C1 and Cf are charged to the input voltage and the input and output terminals of opamp are shorted. Φ_{1d} is another clock which takes transition from 1 to 0 before Φ_1 so that charge injection does not affect the value stored in the capacitor. During Φ_2 C1 capacitor is connected to the output of Sub-DAC. Therefore the output voltage is given by,

$$V_O = \frac{(C_1 + C_F)}{C_F} . V_1 - \frac{C_1}{C_F} V_{ref} \qquad (6.11)$$

In this design C1 and Cf are taken as 0.9pF and 0.3pF respectively.

So, $V_o = 4V_I - 3V_{ref}$

MDAC circuit has been simulated with the input voltage of 70mV and the output of DAC as 167mV (which is the output for 101 sequences) and the simulated results are shown in fig. 6.40.

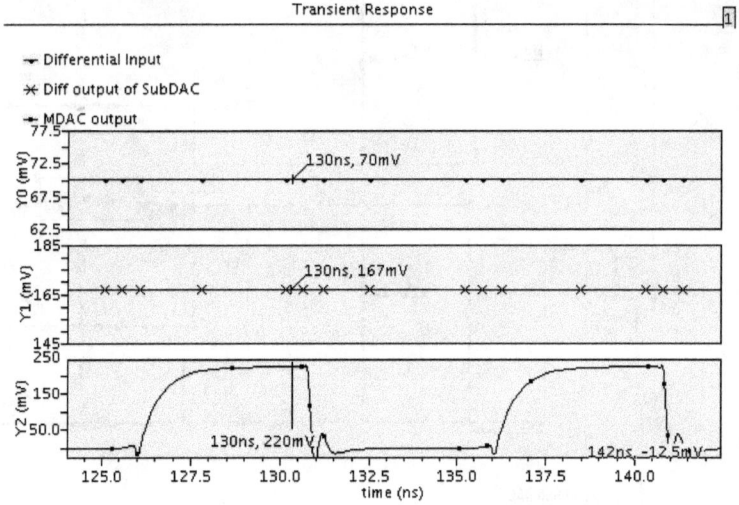

Fig.6.40 Simulated results for the MDAC

6.12 Sub-DAC

The function of the Sub-DAC is to convert the intermediate digital outputs available at every stage into its equivalent analog output that has to be subtracted from the input and passes this residue to the next stage for the conversion. The Block Diagram of the Sub DAC consists of two blocks, one is a DAC encoder and the other is a capacitive circuit.

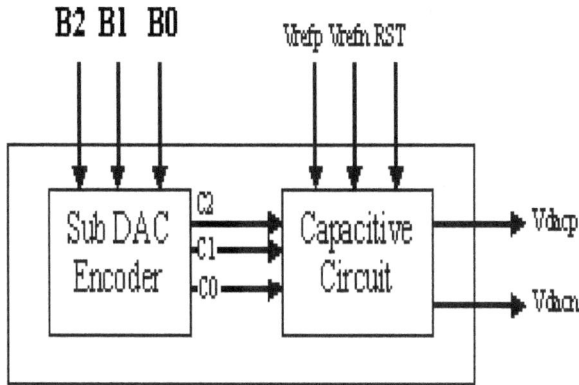

Fig 6.41 Block Diagram of Sub-DAC

Inputs for the Sub DAC are the outputs of Sub ADC (B2, B1, B0), reference voltage (V_{ref}) and Reset Signal (RST). Output of ADC is a 3-bit binary number ranging from 000 to 110. Here in Sub DAC the designer has encoded them into another sequence. The designer encoded the sequence such that the output is 0 for the input of 000. Similarly the other codes and their outputs are observed we can say the outputs of DAC are same for the codes above 011 and below 011 except the sign of the output as shown in table. So in the encoded sequence the designer has used the MSB bit C2 for the sign bit and the remaining bits C1 and C0 to get the required DAC output.

Actual DAC Input B2 B1 B0	Encoded Sequence C2 C1 C0	DAC Output
1 1 0	1 1 1	Vr
1 0 1	1 1 0	2Vr/3
1 0 0	1 0 1	Vr/3
0 1 1	0 0 0	0
0 1 0	0 0 1	-Vr/3
0 0 1	0 1 0	-2Vr/3

152

0 0 0	0 1 1	-Vr

Table 6.6 Input/output Table of Sub-DAC

In this design Vr is chosen as 0.5V. So DAC outputs become 500mV, 333.3mV, 166.7mV, 0, -166.7mV, -333.3mV and -500mV for the encoded sequence of 111, 110, 101, 000, 001, 010 and 011 respectively. To get the required DAC output for the encoded sequence consist capacitors C1, C2 and C3 are in parallel. Capacitor C1 one side consist Voutp another side is input c1. Similarly capacitor C2 one end connected to Voutp and another is c0. Capacitor C3 connected between Voutp and ground. It is half part of the total circuit. It is taken here to explain how sub DAC works. As the complete pipeline ADC is differential in this design, we need Vdacp and Vdacn. Similar to this another half is used on the upper side in the actual implementation for Voutn, as it is a differential circuit. The outputs shown in the table are differential outputs only. C1, C2 and C3 consist of values 2C, C and 3C respectively. Two LSB bits C1 and C0 are the inputs for the two capacitances and the MSB (not shown) C2 is used for the sign. If any bit either C1 or C0 is zero then the capacitors are connected to V_{ref}.

From the fig, if C1=0 and C0=0 then the output is 0, because the capacitors are not charged.

If C1=0 and C0=1 then

$$Voutp = \frac{C2}{C1 + C2 + C3} = \frac{C}{6C} = \frac{1}{6}Vref$$

Similarly

$$Voutn = -\frac{1}{6}Vref$$

It implies Vout = Voutp − Voutn = $\frac{1}{3}$Vref

Similarly outputs can be calculated for the remaining values of C1 and C0 also.

In Sub-DAC we need an encoder to encode the outputs of the sub-ADC output into another sequence.

The logic in the capacitive circuit is in such a way that when C2 is 1 output is of positive magnitude and negative for the 0. The actual implementation of Sub-DAC used in this design is shown in Fig. 6.42.

In actual implementation a RST signal is used to reset the value stored in the capacitor every time to get ready for the next input value. So if RST is logic'0' Sub-DAC output is 0. When RST is logic'1' then the 3-bit digital input is applied to the circuit. i.e., outputs are valid during RST. The capacitor values C1, C2 and C3 used in this design are 0.6pF, 0.3pF and 0.9pF respectively

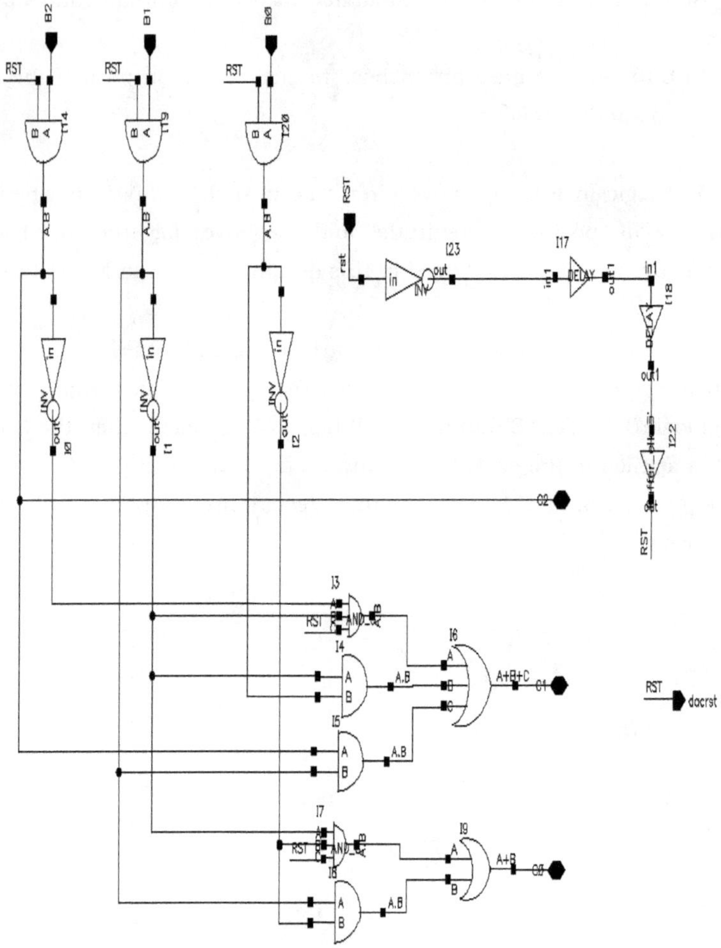

Fig.6.42 Schematic of DAC Encoder

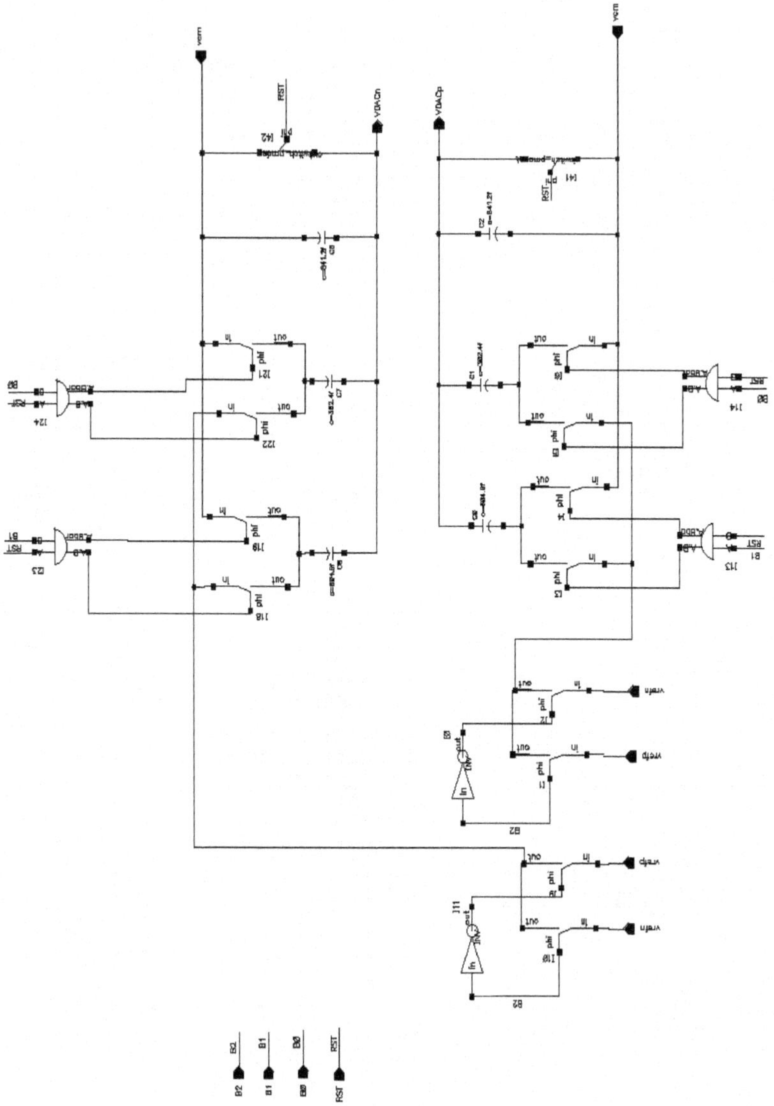

Fig.6.43 Schematic of Sub-DAC

In the Sub-DAC schematic a pmos switch is used in parallel with the capacitor operating on RST. Vcm in this design is 1.65V. So when RST is '0' vdacp and Vdacn are 1.65V i.e., differential output is 0. So the overlap capacitance of switch comes in parallel with the capacitive circuit. So the capacitor C3 becomes 849fF instead of 900fF. The simulated results of Sub-DAC were shown in fig. 6.44 for the input of 101. As shown if RST is 0, output is zero and if RST is 1, output is valid. RST signal here is operated at 100MHz.

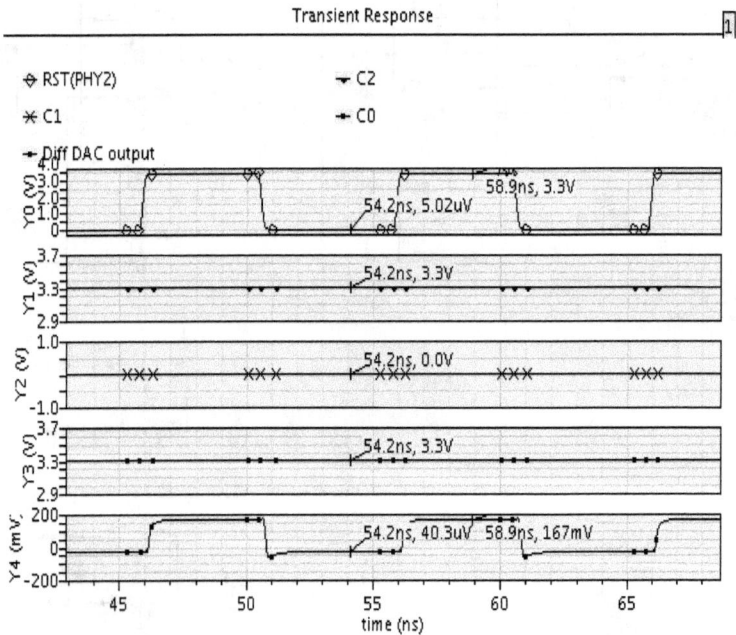

Fig.6.44 Simulated results of Sub-DAC at 100-MHz

6.13 Switch Design

As the pipelined ADC is a switched capacitor circuit, it plays a vital role. At the place of switch we can use a simple NMOS or PMOS or Transmission Gate etc. The schematic of the switch used in this design is shown in fig. 6.45.

In this design a transmission gate with dummy switches on both sides is chosen, as it takes care of the charge injection problem, a major problem in switch capacitor circuits.

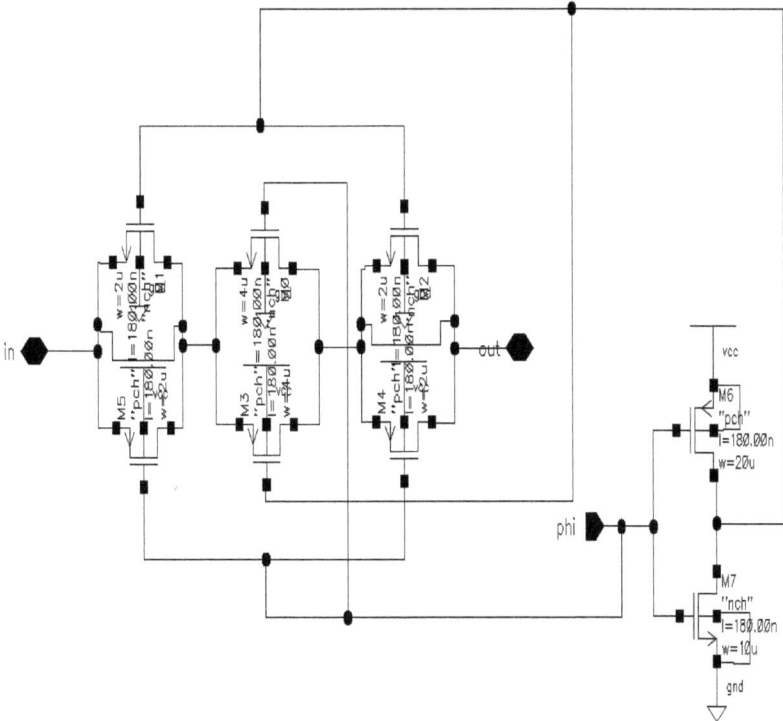

Fig.6.45 Schematic of Switch

But the sizes of the switches used in the entire design are different. At every block the sizes are optimized in such a way that it meets the given requirement. Ultimately they have to operate at 100 M-Hz.

6.14 Buffer Design

In the design at various places buffers are used to increase the drive capability. It is basically an even number of inverters kept in series with increasing aspect ratios (W/Ls).

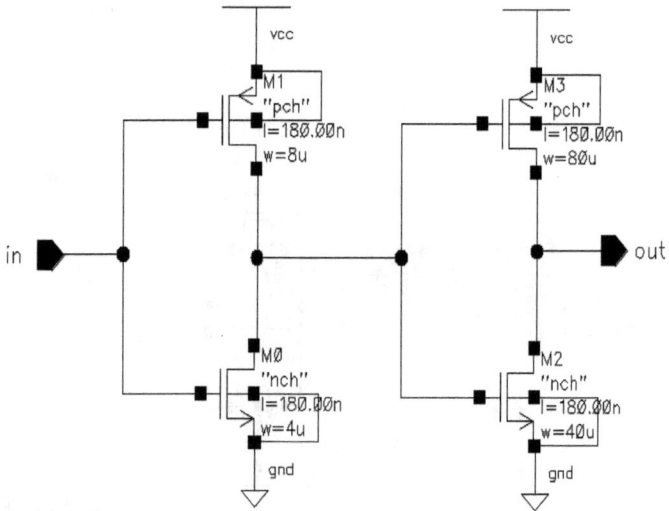

Fig. 6.46 Schematic of Buffer

A simple buffer is shown in Fig. 6.46. The sizes are different as it depends on the drive capability.

6.15 Schematics of Gates used in Design

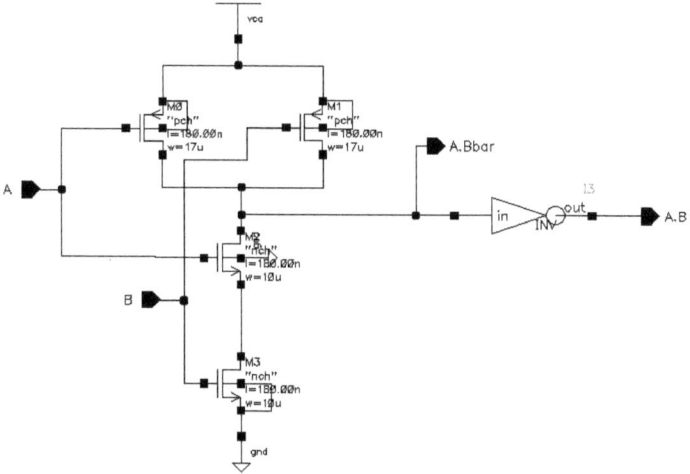

Fig. 6.47 Schematic of 2-input NAND/AND gate

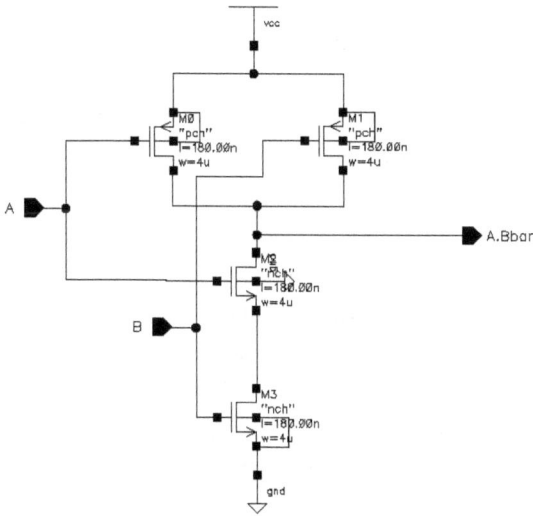

Fig. 6.48 Schematic of AND gate

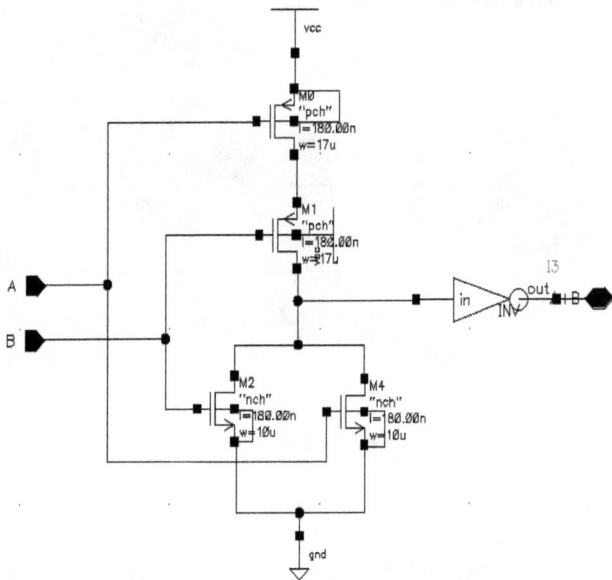

Fig. 6.49 Schematic of OR gate

These are the schematics of Logic Gates used in the design at various places. In this chapter the design of all the blocks used in the pipelined ADC was discussed.

In the previous chapter each block is designed and tested individually. In this chapter integration of the various blocks used in the pipelined ADC is discussed. The problems that we got during the integration are clearly explained with their remedies. Design of the shift register and the Digital Error Correction is explained. The 10-bit pipelined ADC as a whole has been simulated for different values of the input and the results are shown.

6.16 Integration of Sub-ADC and Sub-DAC

For all the odd stages used in pipelined ADC, the comparators latch operates on inverted Φ_{1d}. OE (Output Enable) of Thermometer to Binary code converter operated with some delay of latch because the outputs of comparator take some time to settle. RST in Sub-DAC is operated on Φ_2 which can be seen as the delayed OE. So after the output of sub-ADC is settled then only the RST signal is active. Similarly the even stages operate on opposite phases of odd stages.

6.17 Integration of Sub-DAC and MDAC

In the previous chapter the design of Sub-DAC and MDAC are discussed and they are simulated individually and are working properly at 100M-Hz. But when we integrate the both with the same capacitor values we didn't get the required output. It is an interesting problem called charge sharing. It implies at the output of Sub-DAC we have some capacitance and at the input of MDAC also capacitors are there; so charge has been shared between them. It is clearly explained now: During Φ_1 the capacitors C1 at the MDAC are charged to the input voltage and during Φ_2, C1 at MDAC is connected to the output of Sub-DAC.

So here the charge stored on the capacitor C1 has been shared between C1 and the output capacitance of Sub-DAC. To take care of this problem here a formula has been derived for the capacitor values. Let C_1 be the capacitor at the input of MDAC, C_2 the feedback capacitor of MDAC and C_3 the total capacitance at the output of Sub-DAC. Initially both C_1 and C_2 are charged to V_{in} during Φ_1. During Φ_2, C_3 has been connected to the C_1. So the charge stored on the capacitor C_1 has been shared with C_3.

i.e., $(C_1 + C_3) \times V_2 = C_1 \times V_{in}$

Here V_2 is the voltage on C_1 and C_3 after charge sharing.

So the charge on C_1 becomes $C_1 \times (V_{in} - V_2)$

During $\Phi 1$, C2 is charged to V_{in}. We require another $3V_{in}$ so that output is $4V_{in}$ for the digital input of 000.

$$\frac{C_1(V_{in} - V_2)}{C_2} = 3V_{in}$$

$$=> C_2 = \frac{C_1}{3} - \frac{C_1}{3} \times \frac{V_2}{V_{in}}$$

From

$$\frac{V_2}{V_{in}} = \frac{C_1}{C_1 + C_3}$$

$$=> C_2 = \frac{C_1(V_{in} - V_2)}{3V_{in}}$$

$$=> C_2 = \frac{C_1}{3} - \frac{C_1}{3} \times \frac{C_1}{C_1 + C_3}$$

The 2nd term in the above equation comes into picture due to charge sharing only.

The actual capacitance for C_1, C_2 and C_3 when simulated individually are 0.9pF, 0.3pF and 1.8pF respectively. But when they are integrated C_2 becomes

$$C_2 = \frac{0.9}{3} - \frac{0.9}{3} \times \frac{0.9}{0.9 + 1.8}$$

$$=> C_2 = 0.2pF$$

Here C_3 is the sum of the 3 capacitors used in the Sub-DAC. The 3 capacitors in Sub-DAC are also varied to get the required output. They become 604.9fF, 302.4fF and 841.2fF instead of 600fF, 300fF and 849fF respectively.

By combining the Sub-ADC, Sub-DAC and MDAC one stage of the pipelined ADC has been completed. In this way all the stages are designed and they are integrated. Now the output is a 14-bit digital value which has to be converted to 10-bit using Digital Error Correction.

6.18 Single stage

The complete 2.5 bit stage of a pipelined ADC is shown in fig and the simulation results are shown in Fig. 6.50.

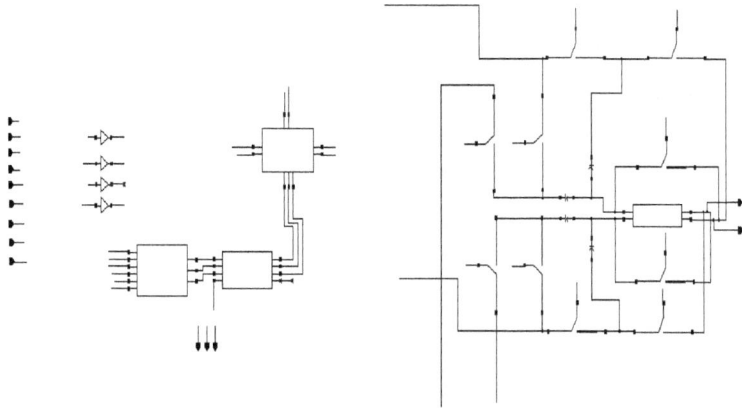

Fig. 6.50 Schematic of Single Stage of pipelined ADC

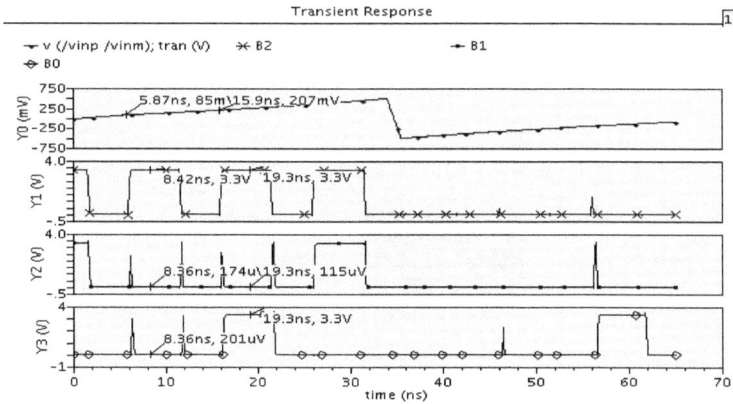

Fig.6.51 Simulation results of a 2.5 bit/Stage at 100-MHz

6.19 Shift Register

In this design the output of the first stage is the input for the 2nd stage and 2nd stage output is the input for 3rd stage and so on. The digital output of the 1st stage is available at the end of Φ_1 and is valid during the Φ_2 period. In Φ_2 only the residue has been applied to the 2nd stage and the 2nd stage digital output is available at the end of Φ_2 and is valid during Φ_1 period. Here as we used 5 stages, the output of 5th stage is available after 5 clock periods from the starting because the odd stages here operate on Φ_1 and even stages on Φ_2. So for the 1st input sample 5th stage output available after 5 clock periods. So designer has to shift the outputs of the every stage in such a way that all the digital outputs for a given sample are available at once. So we have to shift 1st stage output 5 times, 2nd stage 4 times and similarly for later stages as shown in Fig. 6.52.

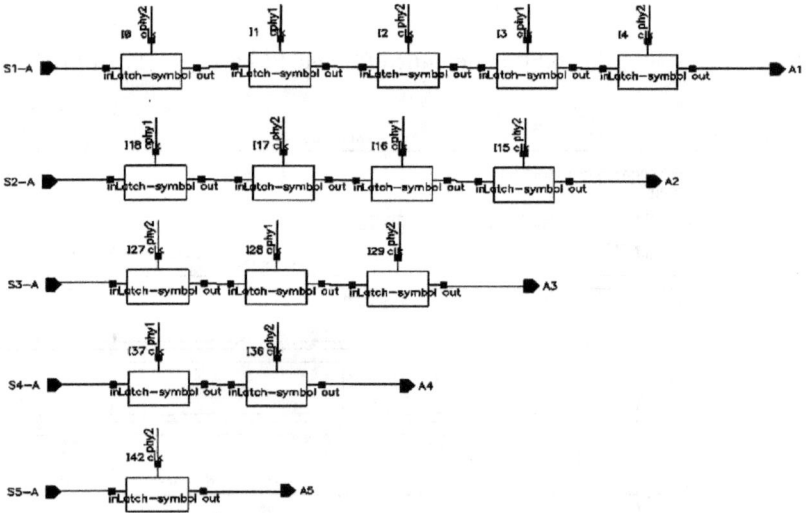

Fig.6.52 Shift Register scheme

From the fig. 6.53 it can be seen that the output for 1st sample available after 5 periods, so it has a latency of 5 periods. S1-A corresponds to output of 1st stage MSB. The basic latch used in the design is shown in Fig. 6.52. In the given figure only one bit of the each stage is shown; in a similar way the 3-bits have to be shifted.

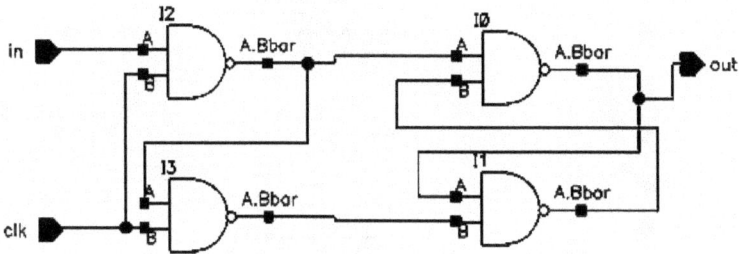

Fig. 6.53 Basic latch used in shift register design

The 14-bit output of the shift register is fed to the digital error correction circuit to obtain the required 10-bit digital output.

6.20 Digital Error Correction Circuit

Digital error correction scheme is one of the good features of the pipeline architecture. Due to this circuitry the requirements on the comparator and the later stages of the pipelined ADC architecture have been relaxed except for the first stage. The scheme behind the digital error correction is already discussed. It simply consists of Half Adders and Full Adders as shown in Fig. 6.54. The complete digital output is available after 5 clock periods i.e., after 3 clock cycles and is called the latency. Once the 10-bit output is available after latency, the output is available for every clock cycle i.e., high throughput rate.

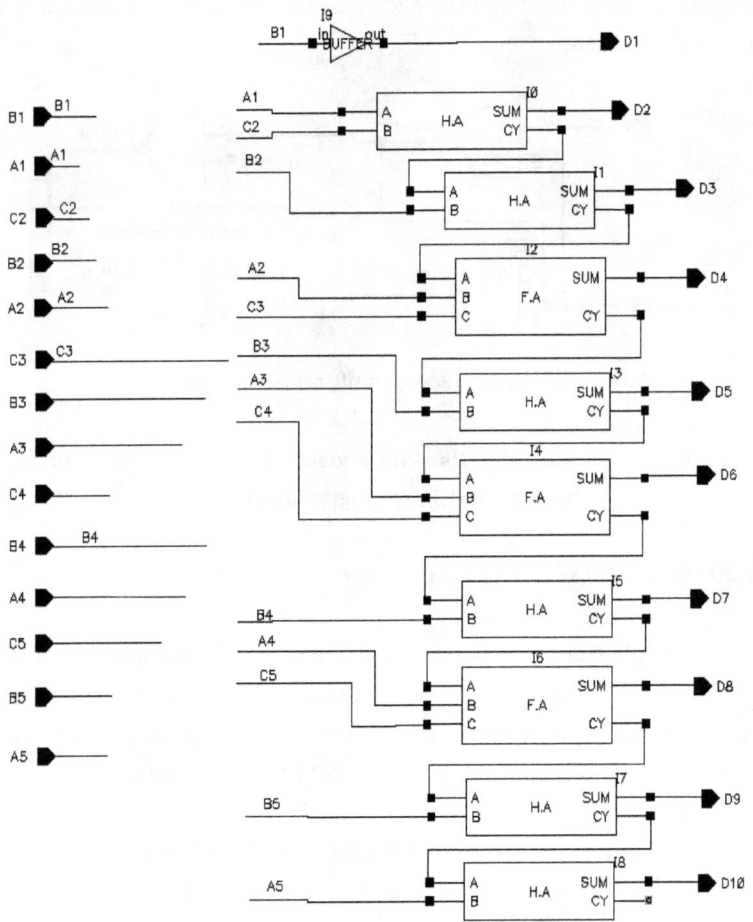

Fig 6.54 Schematic of Digital Error Correction

6.21 10-bit Pipelined ADC

The 10-bit pipelined ADC as a whole with the shift register and Digital Error Correction are shown in fig. 6.55 along with the Test Setup.

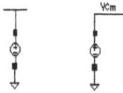

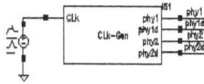

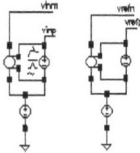

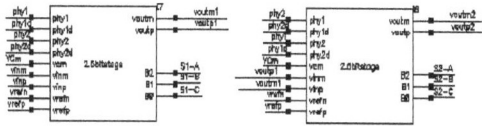

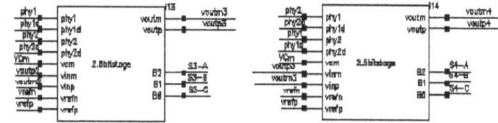

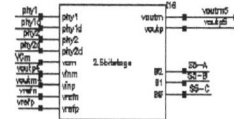

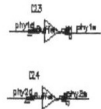

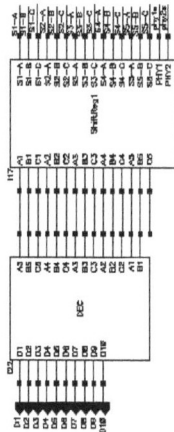

Fig 6.55 Schematic of whole pipelined ADC

168

As shown all odd stages operate on Φ1 and even stages on Φ2. The connectivity in the fig 6.56 is apparent; the connectivity is shown with the wire names in the figure. The circuit has been simulated for various inputs and observed results are correct. Fig 6.56 shows the simulated results for the input of 100mV.

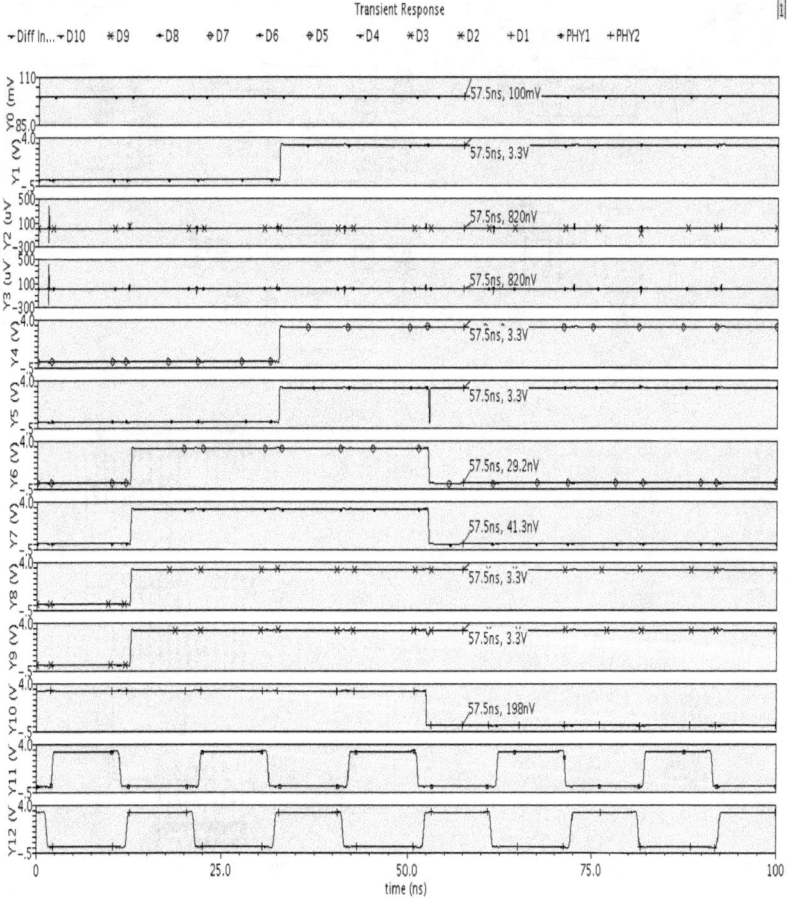

Fig 6.56 Digital output of 10-bit ADC for a 100mV input

As can be seen from Fig 6.57 output is available after 3 Clock cycles, i.e. a latency of 3 clock cycles.

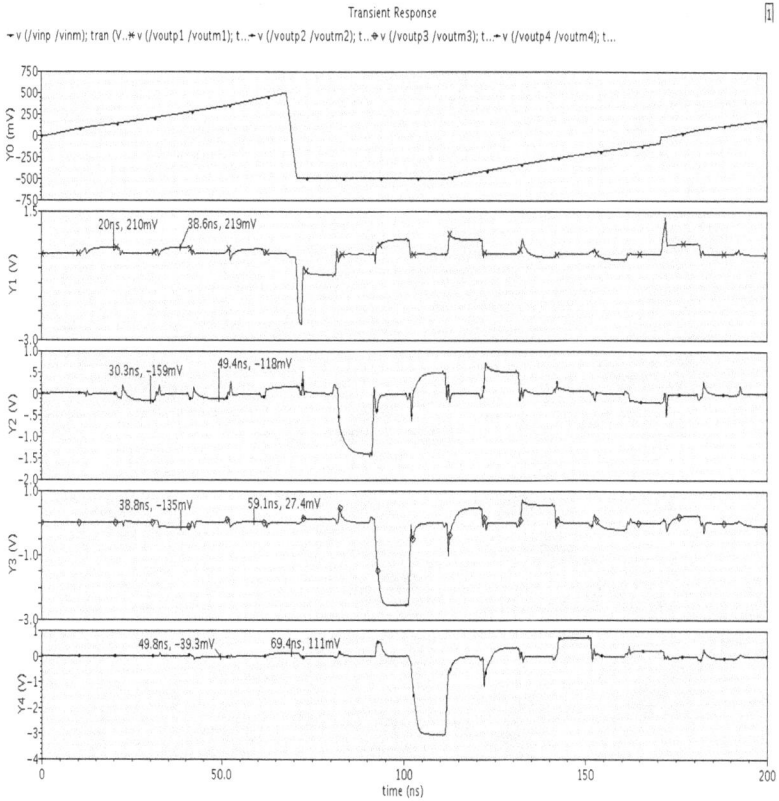

Fig.6.57 Residue Plots at different stages of a pipelined ADC for a RAMP input

Fig. 6.57 shows the output (residue) at every stage of pipelined ADC for a ramp input. As seen from the figure, the 1st stage output available in 1st clock cycle is Φ_1. After that the outputs of remaining stages available in half of the cycle because odd stages operate on Φ_1 and even stages on Φ_2. The error at every stage is measured as less than ½ LSB i.e., less than 0.5mV in this case.

The 10- bit Digital output for a ramp like signal is shown in fig 6.58. In the Figure, D10 to D1 represents the 10 bit digital output with D10 as MSB and D1 as LSB. As can be seen from figure it has a delay of 3 clock cycles. The circuit has been tested for a sinusoidal signal and verified that the results are correct.

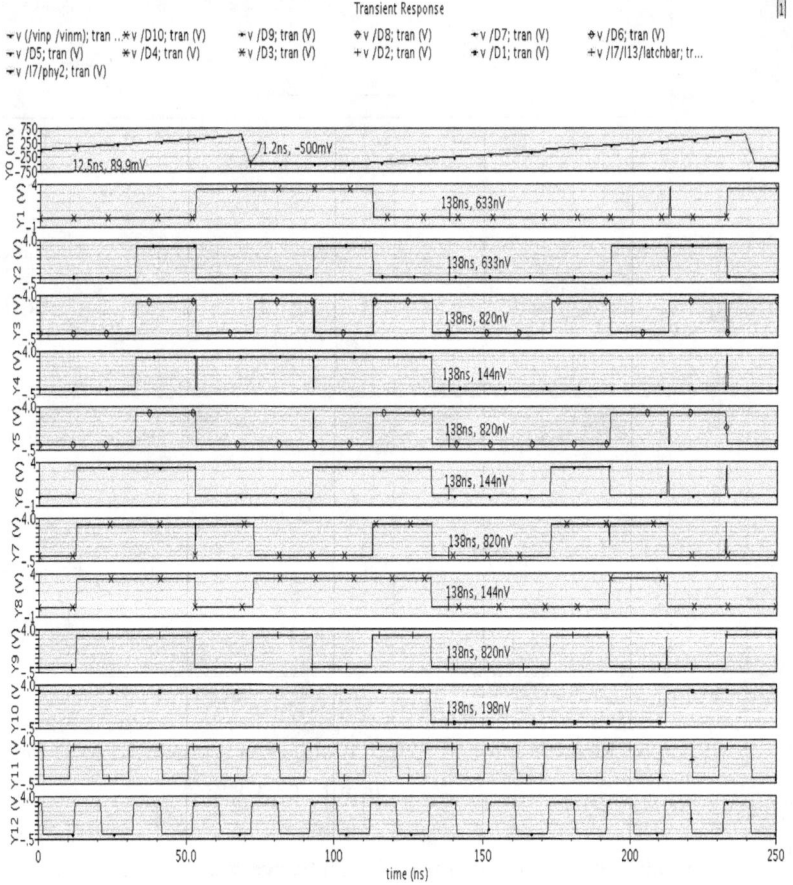

Fig 6.58 Digital output of 10-bit ADC for a RAMP input

The outputs of the individual stages in pipelined ADC for a Ramp input are shown in fig 6.58. The thick lines across the figure show the output of the stages for a particular input. In the Fig 6.59, S1-A, S1-B, S1-C corresponds to the output of the 1st stage. Similarly S5 shows the output of the 5th stage.

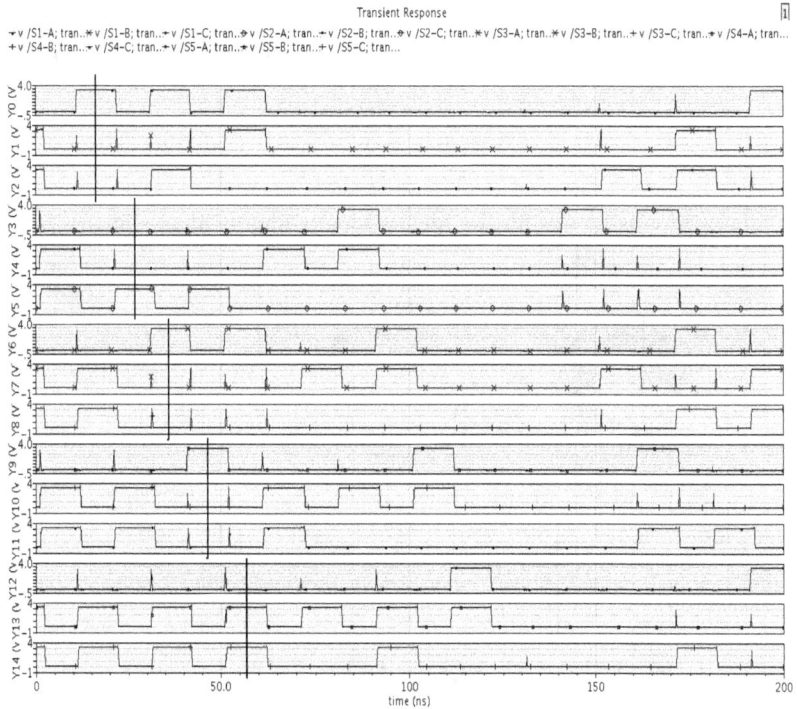

Fig 6.59 Digital output of individual stages for a RAMP input without error

As discussed in previous sections, digital error correction takes care of the comparator offset errors. So here the designer purposefully added an error into the comparator at the 3rd stage of the pipelined ADC, to show that the 10-bit digital output is the correct expected one in spite of the error in the comparator stage. .

Fig 6.60 shows the outputs of the individual stages in pipelined ADC for a Ramp input with error on the comparator at 3rd stage of the pipelined ADC. The thick lines across the figure show the output of the stages for a particular input. The value of the input is 80.5 mV corresponding to an expected ADC output of 1001001001. In the Fig 5.11, S1-A, S1-B, S1-C corresponds to the output of the 1st stage. Similarly S5 shows the output of the 5th stage.

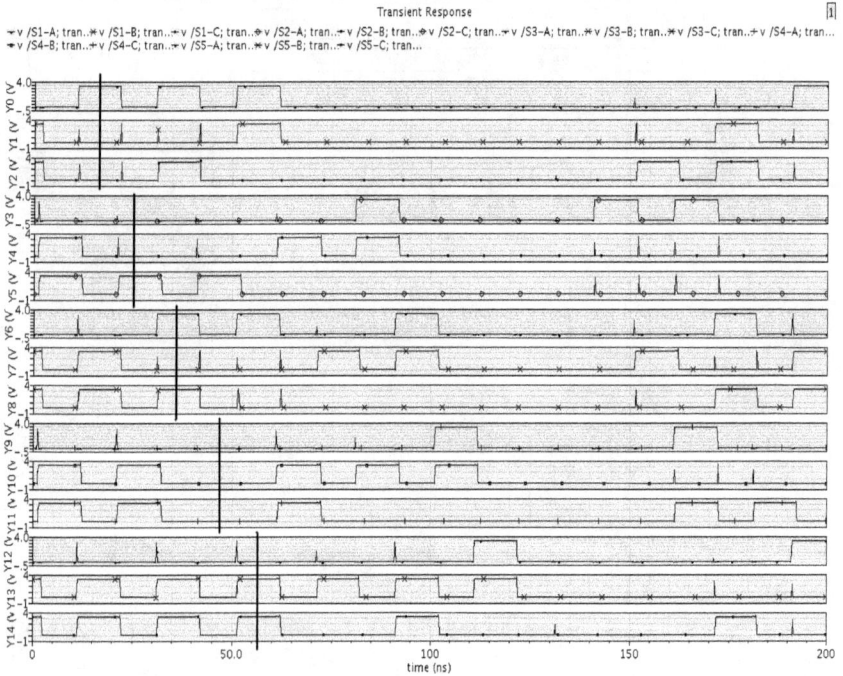

Fig.6.60 Digital output of individual stages for a RAMP input with comparator error

Fig 6.60 shows the 10-bit digital output of the pipelined ADC with comparator error on 3rd stage. The 10-bit output is same as the output without error. This is the advantage of the digital error correction.

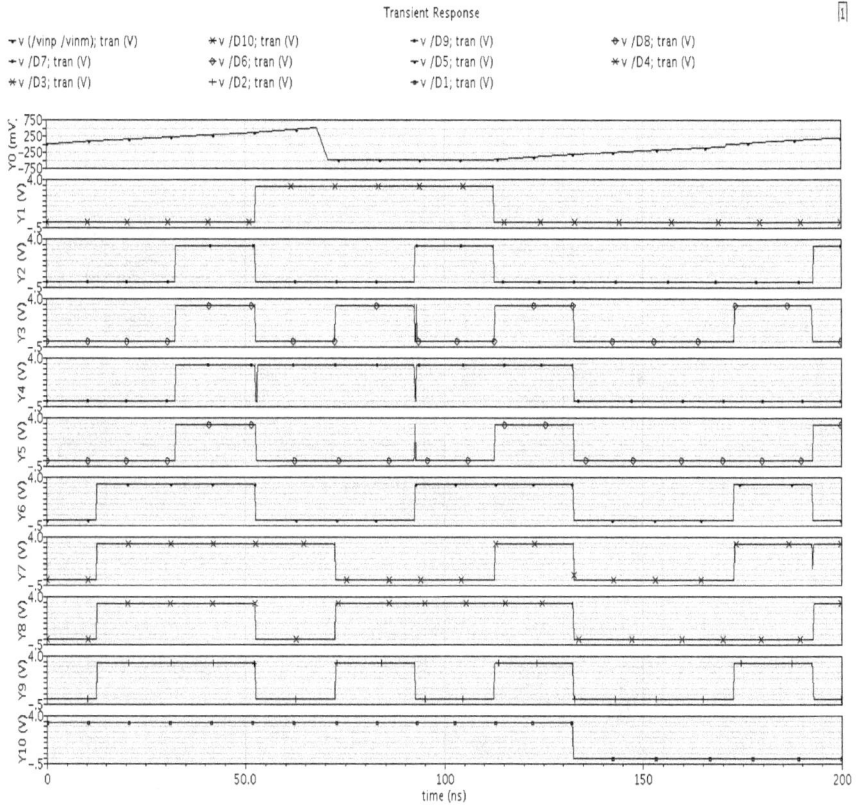

Fig 6.61 Digital output of 10-bit ADC for a RAMP input with comparator error

From the above figures, for the input of 80.5mV the individual stage outputs and final output with and without comparator error are shown in table 6.7.

Stage Outputs	Without Comparator error	With comparator error on 3rd stage
1st stage	100	100
2nd stage	001	001
3rd stage	100	101
4th stage	100	000
5th stage	011	011
Final 10-bit Output	1001001001	1001001001

Table   6.7

Comparison of 10-bit ADC with and without comparator error

As can be seen from table 6.7, the outputs of the individual stages are different for the two cases. But the final 10-bit digital output is same in both the cases, which is the expected output.   This clearly demonstrates the operation of the Digital Error Correction Circuitry. It is the main advantage of the pipelined ADC. It will correct the errors due to comparator offsets and the inter stage amplifier errors.

6.22 Conclusion

In this chapter designed architectures results are discussed. Gain Boosted OTA gain, GBW and slew rate is suitable for considered application. 10 bit A/D application is explained with suitable blocks and waveforms.

CHAPTER 7

CONCLUSION

10-bit, 100MHz pipelined Analog to Digital Converter application designed in 0.18μm CMOS technology. A 2.5 bit /stage architecture is used for pipelined ADC. Whole design was broken into various blocks namely clock generator, Sub-DAC, sub-ADC, gain stage, error correction block and shift register. The specifications of the various blocks in pipelined ADC were derived from the specifications of ADC. As the OTA was central block of the pipelined ADC, the design of ADC was started with the design of OTA. Several architectures were studied and a gain boosted telescopic OTA is chosen. The OTA designed has a gain of 110db and UGF of 1.8GHz with phase margin of 62°.

A Single ended CMOS OTA based on Local Common Mode Feedback (LCMFB) circuit, has been designed and compared with a conventional single ended CMOS OTA. The technique employed leads to significant increase in Gain Bandwidth (GBW), DC gain, Slew Rate (SR) and decrease in the settling time without extra power consumption.

The design technique proposed combines better performance with simplicity of design and suitability for high frequency operation with few modifications on conventational single ended CMOS OTA. It allows not only to avoid limitation on settling time, but also to improve small signal characteristics. Gain boosted OTA provides the suitable gain and high slew rate for analog to digital converters applications. This architecture is more suitable comparing to remaining architectures.

Once the design of OTA was completed it is tested for the closed loop response i.e., for the gain of 4 stage. Then the remaining blocks of ADC such as Clock Generator, Sub-ADC and Sub-DAC were designed. All the individual blocks were designed and optimized to function properly for a clock rate of 100 MHz and then integrated in steps.

During the integration so many problems were encountered. All the problems and their solutions were clearly explained. Then single stage of a pipelined ADC as a whole is tested and verified that it gives required output. After that Shift registers and Digital error correction circuitry were designed. Finally the 10-bit pipelined ADC as a whole with digital error correction circuit was tested for various inputs like constant DC, sinusoidal signal and Ramp signal and observed that the results are correct.

7.1 FURTHER SCOPE OF THE STUDY

177

In analog and mixed signal circuit applications; the accuracy and speed play an important role. The settling behavior of the CMOS op-amps determines the performance of mixed signal circuits. Such as Analog to Digital converters, Sigma-Delta converters, S/H circuits etc. High gain b/w product and single pole assist in quick settling. Capacitors and trans-conductance elements act as major building blocks for various filters and converters.

Further work can be pursed in high gain and high GBW so that circuits can be improved for fast settling behavior in mixed-signal devices. Low frequency applications like hearing aid for deaf the circuit area or design area and manufacturing cost also can be major parameters those can be looked at for improvement. As scaling of analog devices is a road block in improvement, the demand for improvement in design methodology of mixed-signal devices plays an important necessity for future novel applications in high speed and high performance circuits.

www.ingramcontent.com/pod-product-compliance
Lightning Source LLC
Chambersburg PA
CBHW071214210326
41597CB00016B/1814